THE COMPLETE HANDBOOK OF
Maintenance Management

John E. Heintzelman

Prentice-Hall, Inc. / Englewood Cliffs, N.J.

Prentice-Hall International, Inc., *London*
Prentice-Hall of Australia, Pty. Ltd., *Sydney*
Prentice-Hall of Canada, Ltd., *Toronto*
Prentice-Hall of India Private Ltd., *New Delhi*
Prentice-Hall of Japan, Inc., *Tokyo*

Fourth Printing . . . . . June, 1979

Library of Congress Cataloging in Publication Data

Heintzelman, John E
    The complete handbook of maintenance management.

    Includes index.
    1. Plant maintenance. I. Title.
TS192.H44   658.2'02   76-6490
ISBN 0-13-160994-7

Printed in the United States of America

About the Author

John E. Heintzelman is Manager of Facilities Operations Analysis for the Systems Group of TRW Incorporated, where he also serves as divisional consultant in various aspects of maintenance management.

The scope of the author's industrial background spans more than twenty years, during which he has served as a consultant to a variety of companies and governmental agencies in the selection, design, operation and maintenance of facilities. Mr. Heintzelman also has had extensive first-hand experience in designing computerized systems for scheduling, costing, reporting and accounting for maintenance operations and inventory control of maintenance materials.

His formal education includes a BA from Dartmouth College, graduate studies at the University of California at Los Angeles, and a Master's Degree in Business Administration from the University of California at Long Beach. He is a member of the Phi Kappa Phi National Honor Society.

In addition to his professional duties, Mr. Heintzelman serves as instructor in maintenance management in the Adult Education Section at Cerritos College, Norwalk, California. For more than ten years he has been a frequent and popular speaker at meetings and seminars of professional engineering societies and guest lecturer at university classes and training sessions on modern techniques in maintenance management.

To Susan Funk, whose efforts made this book a reality.

What This Handbook Will Do for You

This book is written for the man in charge of maintenance activities. It gives answers for successfully managing maintenance operations, rather than dwelling on the technical or engineering aspects of how to maintain specific types of equipment or facilities.

The book focuses on the five essential management functions of planning, organizing, staffing, controlling and directing. These management responsibilities are applicable regardless of the type of facilities or equipment being maintained. Thus, the methods presented can be used to improve and to reduce the costs of maintenance operations for corporations, schools, hospitals, government agencies or any other organization. The actual examples of application have saved businesses and organizations millions of dollars in cost-effective efficiencies.

TWENTY-FOUR REASONS WHY THIS BOOK WILL HELP YOU SLASH MAINTENANCE COSTS

This book presents twenty-four chapters packed with modern methods to improve your operations and get the most out of each dollar spent on your maintenance activities. It tells you about the proven methods other maintenance managers have used to make their operations more efficient and cost effective. Let's look at the following checklist. This book shows you how to effect lucrative savings when you:

PLAN

- Put together a PLANNING PROGRAM to reduce the costs of your maintenance operations in the short term and over the long run. See CHAPTER 1.
- Establish BUDGETARY CONTROLS on costs and use dollars to quantify plans. See CHAPTER 2.
- Plan MAINTENANCE EFFORT to get maximum efficiency. See CHAPTER 3.
- Properly IDENTIFY WORK that has to be performed. See CHAPTER 4.
- Effectively SCHEDULE WORK. See CHAPTER 5.
- Implement efficient methods of JOB PLANNING AND ESTIMATING. See CHAPTER 6.
- Develop an effective PREVENTIVE OR PERIODIC MAINTENANCE SYSTEM. See CHAPTER 7.
- DESIGN MAINTENANCE INTO FACILITIES in order to reduce operating costs. See CHAPTER 8.

ORGANIZE

- Effectively ORGANIZE A MAINTENANCE DEPARTMENT based upon its particular mission. See CHAPTER 9.
- Use CONTRACTING to get your maintenance program accomplished effectively. See CHAPTER 10.
- CONTRACT FOR CONSTRUCTION PROJECTS. It tells you about the various types of contracts and when to use which type in order to save time and money. See CHAPTER 11.
- CONTRACT FOR JANITORIAL SERVICES so that you get quality performance at minimum cost. See CHAPTER 12.
- CONTRACT FOR GROUNDSKEEPING SERVICES in order to achieve efficiencies and get cost savings. See CHAPTER 13.

STAFF

- Successfully STAFF CRAFTSMEN AND TRAINEES. Determining manpower requirements, using job descriptions, recruiting and selecting candidates, compensating, training and performance appraisal are all covered to provide you with complete information on this part of the management function of staffing. See CHAPTER 14.
- Systematically STAFF SUPERVISION—describe the supervisory positions, pinpoint where to recruit candidates, how to select candidates, and to compensate, train and appraise supervision. See CHAPTER 15.
- Effectively use a full-fledged APPRENTICESHIP TRAINING PROGRAM as part of the staffing function. See CHAPTER 16.

CONTROL

- Establish a CONTROL SYSTEM covering plans and procedures, measuring, reporting, reviewing reports and making decisions. Effective TECHNIQUES FOR CONTROL including schedules, budgets, using labor hours, labor standards, accounting distribution and integration of operations are all covered as means of achieving control and cost efficiencies. See CHAPTER 17.
- Effectively handle WORK RECEPTION by using modern methods to achieve control and cost-effectiveness in receiving requests for work, establishing requirements for periodic maintenance, handling repair work, overhauling and rebuilding equipment and constructing and rehabilitating facilities. See CHAPTER 18.
- AUTHORIZE WORK—in order to control operations and reduce costs. See CHAPTER 19.
- Accent PERFORMANCE ANALYSIS. How to develop a system and how to use performance analysis on budgets, schedules and labor utilization to achieve efficiencies. See CHAPTER 20.
- Apply AUTOMATED INFORMATION SYSTEMS in maintenance operations— the types of systems that have been applied successfully to get information to support the function of control. See CHAPTER 21.

- DEVELOP AUTOMATED INFORMATION SYSTEMS—by following the clear-cut steps and pre-tested criteria to get effective automated systems that will dramatically reduce your operating costs. See CHAPTER 22.

DIRECT

- Inaugurate basic principles that must be applied in effectively DIRECTING MAINTENANCE SUPERVISION. It covers motivating, communicating, leading, teaching, delegating, coordinating and orienting as the necessary prerequisites for proper direction of maintenance supervision. See CHAPTER 23.
- APPLY DIRECTION PRINCIPLES—how maintenance managers have used staff meetings, problem-working sessions, touring, coaching, the principle of exception and the doctrine of completed staff work to effectively execute the management function of direction and get the most out of their supervisors. See CHAPTER 24.

HOW THIS BOOK IS ORGANIZED

The Complete Handbook of Maintenance Management is organized into five sections. Each section covers one of the five basic management functions of planning, organizing, staffing, controlling and directing. Within each section there are two or more chapters about methods of performing the management function. Although these basic functions are applicable to any manager, the book is tailored to their effective application by the manager of maintenance activities.

HOW TO USE THIS BOOK

You can apply this book in many ways, but here is a recommended nine-step procedure that will put the five basic functions of a manager into action.

1. Skim over the entire book to orient yourself to its contents.
2. Read the section on the basic management function you first want to analyze and improve in your operation.
3. Go over the section again and list the methods you believe are germane to your maintenance activities.
4. Orient your supervisors and immediate superior as to what you intend to do. (You might even encourage them to also read the appropriate section of the book. Better still, have them get their own copy of the book. This will help them, you, the publisher and the author.)
5. Plan with your supervisors how to implement the selected methods.
6. Select the people who will be responsible for implementing the methods.
7. Assign the necessary tasks and responsibilities in writing.
8. Advise your supervisors and immediate superior as to who these people are and what they have been assigned to do.
9. Follow up to assure events are conforming to plans.

Now, if you want to find out why these nine recommended steps in the "action program" of a productive maintenance management system really work, read this book. Between its covers are the practices and procedures, replete with forms, checklists and other working tool aids, for performing at the highest peak of professionalism and attaining enviable bottom-line results.

John E. Heintzelman

CONTENTS

ILLUSTRATIONS

16

SECTION I—PLANNING

Food for thought:

A dull axe requires great strength; be wise and plan when to sharpen the blade.

A prudent man foresees the difficulties ahead and prepares for them; the simpleton goes blindly on and suffers the consequences.

The wise man looks ahead. The fool attempts to fool himself and won't face facts.

There are no problems—only challenges and opportunities.

POINTS TO CONSIDER

Planning involves looking ahead and making decisions on what you are going to do. When you have no solid plans of your own, you react to crises rather than grabbing initiatives. Planning is what permits you to act rather than react. It permits you to sell your ideas to upper management and assures that your maintenance activities are cost-effective over the long run.

This section has nine chapters describing how to accomplish the management function of planning in a maintenance department. The first chapter gives you some guidelines. The next chapter covers the use of dollars as a means of quantifying plans. The following four chapters deal with: planning the level of maintenance effort, identifying and scheduling work, job planning, and job estimating. Chapter 7 tells how to develop an effective periodic maintenance system. The last chapter covers planning maintenance in new facilities.

Without planning, you will never get from here to there in effectively organizing, staffing, controlling or directing your maintenance operations.

1 Setting Maintenance Goals, Policies, Programs and Procedures

Planning involves the selection of objectives and the determination of the policies, programs and procedures to be used for achievement of the selected objectives. Because it involves selection among alternatives, planning is decision making. Of all the management functions, planning is the one that permits a maintenance department to act rather than react. When given proper attention, it is the function that facilitates the maximum utilization of available labor, money and material resources. Planning, however, is the function that often receives the least emphasis. This lack of emphasis causes much of the criticism leveled at maintenance departments for being too costly or unresponsive to the organization's needs. Without conscious goals, policies, programs and procedures, execution of the planning, organizing, staffing, controlling and directing functions will be suboptimal.

Effective plans do not just happen. It requires judicious thought to determine what type of planning should be done, who should do it and when and how it is to be done.

How the maintenance department is organized to achieve its mission also requires planning. The organization structure should be continually appraised in the light of revisions of department objectives, changes in the organization serviced by the maintenance department and external factors that impinge on maintenance operations.

Staffing involves manning, and keeping manned, the positions required in the organization structure. This function includes defining manpower requirements, recruiting, selecting and hiring candidates, compensating employees, and training and developing employees to accomplish their tasks. Staffing cannot be effectively accomplished if you do not plan how you are going to get it done.

Directing involves supervising and guiding subordinates, that is, orienting and motivating them, guiding them toward improved performance, and clarifying their assignments. If you do not have programs and procedures to accomplish these tasks, your chances of doing a good job of directing are pretty slim.

Planning the execution of the control function is also essential. You cannot measure performance, correct negative deviations, assure the accomplishment of plans or make necessary revisions to plans without first determining what techniques you are going to use and how they are to be applied.

Some scholars and academicians concern themselves with the order in which the management functions are accomplished. In theory, planning comes first, and organizing, staffing, directing and controlling follow in sequence. According to this logic a maintenance department carries out only one, all-encompassing master plan. In the real world, however, a maintenance manager may find himself performing all his functions at the same time. Plans lead to subordinate plans, existing plans require changes and new plans are developed while old ones are being executed. Accordingly, it is impractical to insist on any special time sequence for the various functions except to say that planning should precede the performance of any of the others. One maintenance manager claims that staffing is the most important planning he does. With the right people, all the rest of it is easy.

This chapter provides proven guidelines that have enabled actual companies to plan on a systemized basis, realizing dramatic savings in the process. To the seasoned maintenance manager a portion of this material may be so obvious that he need only gloss over it. For other readers it is hoped that this presentation will shed some light on problem areas.

THE FIVE-TO-ONE MULTIPLIER

There is a rule of thumb that says: for every dollar you spend in proper maintenance planning you will save at least five dollars in subsequent expenditures. The validity of this rule is demonstrated by the following examples.

A large amusement park located in Southern California created a maintenance planning staff. The costs of the new organization amounted to around $58,000 a year for payroll and office expense. The results of their efforts in planning operations reduced annual maintenance costs by $300,000 in their first nine months of existence.

For a new airport in the Midwest, a consulting firm was paid $35,000 to do a planning and budgetary cost-estimating study for the operation and maintenance of certain terminal buildings. The participating airlines followed the consultants' recommendations regarding specifications for bidding janitorial work and jobs such as ramp sweeping, window washing, and man-loading of the in-house maintenance crew. For one airline the resulting annual cost for total operations and maintenance amounted to $4.81 per square foot. For a similar terminal building that was not included in the study the cost was $5.88 per square foot. The difference of $1.07 per square foot saved the airline that participated in the study over $700,000. Other participating airlines experienced a cost of around $4.38 per square foot. For each of them the difference of $1.50 per square foot saved more than $750,000 annually. Thus, the $35,000 investment the participating airlines made in the planning contract yielded an annual savings of over one million dollars in the first year of operation in their new facilities.

THE SPANS OF PLANS

Plans are sometimes classified based upon the period of time they have been designed to cover. Long-range plans normally project three to ten years into the future. Short-range plans normally cover a time span of one to three years. Near-term plans cover monthly and quarterly time periods.

LONG-RANGE PLANS

Some maintenance managers claim that long-range plans are impossible to develop, because they just cannot correlate meaningful information. The manager who makes this claim is generally the one who also seems to muddle from crisis to crisis. Long-range plans can be meaningful, provided they include elements or objectives that can be predicted with a reasonable amount of certainty such as:

- Painting of buildings
- Rehabilitation or replacement of roofing
- Rehabilitation or replacement of building equipment such as hot water heaters, boilers, or air conditioning compressors
- Replacement of maintenance department motor vehicles
- Replacement of maintenance department shop equipment
- Retirement and replacement of permanent maintenance department employees

The requirements of long-range plans should be communicated to other departments involved in the process of their implementation. For example, if the comptroller is not aware of capital requirements for rehabilitation of facilities, he can hardly be expected to include funding for such items in his long-range fiscal plans. Without his support and the inclusion of the necessary funding in his plans, the maintenance department cannot expect to get its long-range plans implemented.

LONG-RANGE PLANS EVOLVE TO THE PRESENT

As time passes, long-range plan objectives become part of short-range plans and, ultimately, near-term operating plans. Progression of this sequence may take up to ten years. When a planned objective has been considered and evaluated over such a great amount of time, its validity certainly has a better opportunity for testing than if the item had been initiated on a crash basis by means of a supplementary budget. Objectives determined on a crisis basis can seldom have this validity, because there is insufficient time in which to properly evaluate the various alternatives.

A classic example of the consequences of crisis planning is shown by the case of the manufacturing plant that was constructed on a crash basis on the East Coast. In order to expedite the project the Southern California parent company used a West Coast architectural firm to prepare the plans and specifications, including landscaping plans. Used to Southern California requirements for heavy watering of lawns, the architect included the installation of a sprinkler system in the lawn areas. The site where the plant was constructed normally had sufficient rainfall in the summer months and such watering was not necessary. Thus, for five years the sprinkler system lay dormant. Finally, there was a drought one summer and lawn watering was required for the first time. By then, the sprinkler heads had all been buried. They had to be fixed along with some piping and valves before the system would function properly. The cost of rendering the system operable was four times the cost of inexpensive hoses that would have provided the watering necessary during the unusual drought. Thus, the original crash decision to install a sprinkler system in the first place was compounded by another crisis decision to refurbish the sprinkler system to meet the rare occurrence of a drought period.

ESTABLISHING SHORT-RANGE PLANS

Short-range plans are generally easier to develop than long-range plans. There are fewer uncertainties to contend with because the time frame is shorter. Thus, the planning process is facilitated. There was a time when short-range plans covered a period of up to five years. This practice has been largely abandoned because specific objectives in the fourth and fifth years could not be developed with the precision of a plan covering only three years.

Normally short-range plans should contain considerably more detail for the first year covered than for succeeding years. The last two years of the plan are more generalized, showing categories of effort and major programs to be started, accomplished or completed. The first year of the short-range plan should interface directly with the established budget for the year.

Figure 1-1 is an example of a short-range plan sheet for painting operations covering three years. Figure 1-2 is an example of the more detailed presentation of part of the first year of the plan. Note that dollar values are rounded for budgetary and planning (B&P) purposes to $500 increments in Figure 1-1 and to $100 increments in Figure 1-2. These planning documents are not intended to indicate precise dollar and cent figures.

MAINTENANCE THREE-YEAR OPERATING PLAN			
		FISCAL YEARS	
TYPE OF WORK	FIRST	SECOND	THIRD
Interior Painting of:			
Public/Service Areas	$ 48,000	$ 50,400	$ 53,000
Offices	37,500	39,500	41,500
Exterior Painting of:			
Roof Equipment	25,000	26,500	28,000
Buildings and Grounds	80,000	84,000	88,500
TOTALS	$190,500	$200,400	$211,000

FIGURE 1-1

SAMPLE SHORT-RANGE PLAN SHEET

**19XX
MAINTENANCE OPERATING PLAN
INTERIOR PAINTING—PUBLIC/SERVICE AREAS**

Building Number	Budgetary and Planning Cost Estimate
R1	$ 2,400
E2	4,100
M1	1,400
M2	1,500
M3	3,300
M4	2,200
O1	2,700
R5	4,100
R4	2,500
R6	6,300
M5	5,800
R3	2,000
S	5,100
E1	2,900
R2	1,700
	$48,000

FIGURE 1-2

SAMPLE DETAILED FIRST YEAR OF SHORT-RANGE PLAN SHEET

NEAR-TERM PLANS

Near-term plans contain the most detail and represent the operating plan for each month or week in the period covered. These plans contain the milestone dates for each construction or modification job being performed and the number of man-hours allocated to breakdown maintenance, repair or rehabilitation effort, preventive maintenance, and so on. Thus, the near-term plan constitutes an allocation of resources to the various functions or types of work to be performed by the maintenance department. The plan should constitute the compilation of all work to be performed.

Where detailed estimates have been prepared for work, the hours estimated for each craft skill provide the data for developing the near-term plan for the period during which the work is to be performed. Figure 1-3 is a sample of a quarterly operating plan for the interior painting of the public/service areas indicated in Figure 1-2. Note that the work is now expressed in terms of standard labor hours in lieu of dollars with a specific performance schedule in terms of weeks.

Allocation of man-hours for breakdown maintenance requires predictions: first, the number of trouble calls the maintenance department can expect to receive; second, the number of man-hours in each craft skill that will be required to respond to these calls and restore operations.

One method of prediction is to use a simple projection of trends by craft skill based upon the activity of the previous month or months. A somewhat more sophisticated method is to project certain types of failures based upon the season of the year. For example, during the summer months there should not be any failures in a steam heating system, since it is not being operated. Accordingly, no man-hours should be allocated to this type of breakdown work during certain months of the year. A simple projection of trends would not yield the same results, since it would not provide for such seasonal adjustments. Any projection-of-trends method of forecasting is predicated upon the assumption that history will repeat itself. Thus, any forecast using a projection-of-trends method is only as useful as the validity of this assumption.

MAINTENANCE SCHEDULE
19XX—FIRST QUARTER
INTERIOR PAINTING—PUBLIC/SERVICE AREAS

Building Number	Standard Labor Hours	Weeks Required	Cumulative Weeks Required	Scheduled Performance Weeks
R1	400	2.0	2.0	01-03
E2	680	3.4	5.4	03-06
M1	220	2.2	7.6	06-08
M2	245	2.5	10.1	08-11
M3	540	2.7	12.8	11-13

FIGURE 1-3
SAMPLE NEAR-TERM PLAN SHEET

USING TIME AS A DIMENSION OF PLANNING

Calendar time is a unit of measure essential to proper planning. One of the most common planning tools is the budget, which is expressed in terms of time as well as dollars.

If plans are not expressed in terms of time, they are merely dreams. Target dates are essential to the accomplishment of jobs. Employees should always know when they are expected to perform a task. Without the assignment of specific target dates they cannot be held properly accountable for performance. There is no equity in chastising an employee for failing to perform a task if you have failed to tell him when you expected the job to be done.

By nature, human beings tend to solve the immediate problems of the day and to put off the others until tomorrow. To effectively overcome this phenomenon, dates must be set for the execution of plans. Assignment of priorities is then forced, and the time needed for completion of plans can be measured against scheduled dates.

Situations in which events are not conforming to plans can thus be easily discovered, allowing for adjustment of plans based on the most recent information available. Plans must be flexible. When they are not they lack realism and can become ends in themselves rather than means to an end.

SCHEDULING

Scheduling is a method of planning that cites specific objectives to be accomplished in relation to time.

Schedules can be applied to a large number of activities in a maintenance department. Planning of rehabilitation and modification jobs assigned to the maintenance department for execution should be scheduled. Inherent in any preventive maintenance program is the use of schedules to delineate when specific types of maintenance are to be performed. Near-term operating plans should also be reduced to specific schedules with certain weeks or months set aside in which specific jobs should be accomplished.

Until an annual budget has been received, scheduling on major construction projects is best done in terms of estimated time periods expressed in days or weeks. After receipt of the annual budget, the estimated time schedule should be converted to specific calendar dates. When a schedule uses calendar dates prior to receipt of an approved budget, the dates are often rendered obsolete; if the budget is received after the scheduled start date, the schedule must be totally reworked to start when the authorization to proceed is received. Rescheduling is not necessary if calendar dates are not assigned to the schedule until after an annual budget is approved. Figure 1-4 is an example of a schedule expressed in estimated blocks of time. Figure 1-5 is an example of the same schedule to which calendar time has been applied after receipt of the annual budget. Note that the time for preparation and approval of the job order took eight weeks rather than the anticipated four weeks, due to a four-week delay in receiving an approved annual budget.

The use of schedules as a planning tool is not restricted to internal operations. Schedules may also be used as a communication and coordination device in dealing with other departments. Examples of scheduling that may be used with other departments follow:

- The renewal of dates required for service contracts or blanket purchase orders can be submitted to the Purchasing Department to help them plan their workload.
- The renewal dates for maintenance vehicle registrations, elevator permits, etc. can be submitted to the Accounting Department to notify Accounts Payable of these requirements for disbursements prior to the specific expiration dates.
- The planned dates for painting offices can be coordinated with the people occupying the space.
- The replacement dates for retiring employees can be submitted to the Personnel Department to help them plan recruitment and retirement activities. (This has a direct relationship to performance of the staffing function.)

PROJECT MASTER SCHEDULE

PROJECT TITLE: Restroom Modifications

FUNCTIONAL AREA: Manufacturing—Final Assembly

LOCATION: Building 2 Northwest Corner—1st Floor

DATE PREPARED: 1 February 19XX

PREPARED BY: Jay Funk
Maintenance Planning

TASK DESCRIPTION	WEEKS REQUIRED	CALENDAR DATES	REMARKS
1. Job Order Preparation & Approval	4		
2. Design	4		
3. Design Review & Changes	1		
4. Detailed Cost Estimate	1		
5. Plan Check & Permits	3		
6. Release of Work Order & Drawings	1		
7. Demolition	2		
8. Construction	8		
9. Inspection & Punch List	1		
10. Punch List Work Off & Final Completion	2		
TOTAL ESTIMATED TIME	27		

FIGURE 1-4

SAMPLE PROJECT SCHEDULE USING ESTIMATED TIME
PENDING ANNUAL BUDGET APPROVAL

PROJECT SCHEDULE

PROJECT TITLE: Restroom Modifications

FUNCTIONAL AREA: Manufacturing—Final Assembly

LOCATION: Building 2 Northwest Corner—1st Floor

DATE PREPARED: 1 February 19XX

PREPARED BY: Jay Funk
Maintenance Planning

DESCRIPTION	WEEKS REQUIRED	CALENDAR DATES 19XX	REMARKS
1. Job Order Preparation & Approval	8	1 Feb. to 28 March	4-week delay in annual budget
2. Design	4	31 March to 25 April	
3. Design Review & Changes	1	28 April to 2 May	
4. Detailed Cost Estimate	1	5 May to 9 May	
5. Plan Check & Permits	3	12 May to 30 May	
6. Release of Work Order & Drawings	1	2 June to 6 June	
7. Demolition	2	9 June to 20 June	
8. Construction	8	23 June to 15 August	
9. Final Inspection & Punch List	1	18 August to 22 August	
10. Punch List Work Off & Final Completion	2	25 August to 5 Sept.	
TOTAL ESTIMATED TIME	31		

FIGURE 1-5

ESTIMATED PROJECT SCHEDULE WITH CALENDAR DATES
APPLIED AFTER BUDGET APPROVAL

KEEPING TRACK OF PLANS

Some large maintenance departments prepare a number of various plans and have found that a planning calendar is a useful tool. It facilitates the timing of the preparation of each individual plan to permit proper integration with other plans. The calendar indicates the development schedule of each of the various plans including start date, review dates, and completion date. The format of the calendar may be a box chart or a milestone graph. Figure 1-6 is an example of a planning calendar page using a box chart format. Note that the item covered is the Capital Equipment Plan for the annual budget; this would be scheduled in the planning calendar for integration with the other components of the annual budget.

PLANNING CALENDAR	ITEM—CAPITAL EQUIPMENT PLAN FOR ANNUAL BUDGET	SCHEDULE WEEKS			
RESPONSIBLE PERSON	DESCRIPTION OF ITEM	DEVELOP		REVIEW	FINALIZE
		START	COMPLETE		
Carpenter Foreman	Capital Equipment Requirements List	1	4	5	6
Painting Foreman	Capital Equipment Requirements List	1	4	5	6
Mechanical Foreman	Capital Equipment Requirements List	1	4	5	6
Maintenance Office	Capital Equipment Plan	4	6	7	8

FIGURE 1-6

SAMPLE BOX CHART FORMAT OF A PLANNING CALENDAR PAGE

QUANTIFYING PLANS

Unless plans are defined in terms of numbers as well as time, they will probably remain dreams. Effective planning necessitates quantification.

Mathematics is the queen of the sciences. Modern techniques of administration continue to rely more and more on quantified data. Today, no maintenance department can expect to perform effectively without quantifying what it is doing and what it intends to do in the future. These future intentions are encapsulated in plans.

Dollars is the language of top management. Therefore, to most effectively communicate with top management, the plans of a maintenance department should be expressed and quantified in terms of dollars. The entire second chapter of this book is devoted to the use of dollars in planning.

There are, however, other methods of quantification that can be used by a maintenance department to express plans. Some of these units of measure are headcount, labor hours and premium-time labor hours. These planning techniques facilitate the execution of both the staffing and controlling functions. They are units of measure generally more useful for plans used internally by a maintenance department, particularly those involving staffing and controlling in-house labor. You hire and train people, not dollars. The maintenance department consists of a headcount in various craft skills and work is measured in terms of labor hours. Thus, the dollar unit of measure is appropriate for expressing plans to top management, but it is not the only unit of measure that should be applied in planning.

HEADCOUNT PLANNING

Basically there are two methods for planning headcount. One approach is called MACRO the other MICRO.

USING MACRO HEADCOUNT PLANNING

The MACRO approach consists of estimating required manpower in gross terms based on rules of thumb or industrial standards. Examples are:

- One carpenter is required for the architectural maintenance of 300,000 square feet of office space.
- One janitor is required for the janitorial maintenance of 12,000 square feet of office space.
- One janitor is required for the janitorial maintenance of 18,000 square feet of general industrial area consisting of offices and manufacturing area.

Applying the first rule of thumb listed above to a plant of 600,000 square feet would yield a headcount of two full-time carpenters for the purpose of performing architectural maintenance. (600,000 ÷ 300,000 = 2)

The primary advantage of the MACRO method is that it provides a rapid way of computing headcount requirements. The use of broad standards is sometimes criticized as being too generalized to have validity in the specific situation to which it is being applied.

USING MICRO HEADCOUNT PLANNING

The MICRO method of planning headcount requires substantially more effort than does the MACRO method. It normally consists of the following four steps.

1. The application of time standards to specifically identified work elements to be performed by a given craft
2. The summation of the hours required for each work element to arrive at the total standard hours for the craft skill
3. The computation of the net annual straight-time hours available for performance of work by a full-time employee
4. The conversion of total standard hours required to a headcount figure

As a simplified example of the MICRO method, let us use a manufacturing plant with machine tools. Of these machines, 100 require a weekly oiling operation and

955 of the machines also require a more extensive annual lubrication. In addition, 375 of the machines require that the sludge be pumped out of the coolant tanks every three months.

Weekly oiling, annual lubrication and coolant tank cleaning are all operations that can be performed by an individual classified as a machine tool oiler. These operations do not require the skill level of a machine tool mechanic.

For each of these three operations, two alternate time standards are assignable. As part of step 1 each machine tool in the inventory has been analyzed and one of the time standards has been assigned for each operation applicable to the machine.

The computations that would then be used to arrive at the headcount requirement for machine tool oilers are as follows:

STEP 1—Compiling Standard Times for Each Operation

1. Weekly Oiling (100 machines total)

    a. Standard of three-tenths of an hour per machine (50 machines)
        50 machines x .3 = 15 hours
        per week x 52 = 780 hours annually

    b. Standard of two-tenths of an hour per machine (50 machines)
        50 machines x .2 = 10 hours
        per week x 52 = 520 hours annually

    c. Total Annual Hours = 1300 (780 + 520)

2. Quarterly Tank Cleaning (375 machines total)

    a. Standard of 1.5 hours per machine (75 machines)
        75 machines x 1.5 hours = 112.5 hours
        per quarter x 4 = 450 hours annually

    b. Standard of one hour per machine (300 machines)
        300 machines @ one hour each requires
        300 hours per quarter x 4 = 1200 hours annually

    c. Total Annual Hours = 1650 (450 + 1200)

3. Annual lubrication (955 machines total)

    a. Standard of one hour per machine (500 machines)
        500 machines x one hour = 500 hours annually

    b. Standard of eight-tenths of an hour per machine (455 machines)
        455 machines x .8 = 364 hours annually

    c. Total Annual Hours = 864 (500 + 364)

STEP 2—Summation of Required Hours

Operation	Annual Hours
Weekly Oiling	1300
Quarterly Tank Cleaning	1650
Annual Lubrication	864
Total Hours Required	3814

STEP 3—Computing the Net Available Hours

Annual Gross Available Hours = (40 hours/week x 52)		2080

Deductions:

Vacation (10 days)	80	
Sick Leave (5 days)	40	
Holidays (10 days)	80	200
Net Annual Available Hours for a Full-Time Employee		1880

STEP 4—Conversion of Required Hours to Headcount

Standard Required Hours ÷ Net Available Hours = Required Headcount
    3814 ÷ 1880 = 2.02 or 2 Machine Tool Oilers

The figure of 2.02 computed above in step 4 has been rounded to arrive at the required headcount of two full-time machine tool oilers. During the year the two oilers would provide 3760 hours (2 x 1880) for performance of the 3814 standard hours of work. The annual variance would be 54 standard hours (3814 less 3760). In essence, this means that the two machine oilers would be overloaded by 54 hours in the performance of a year's work. Since 54 hours represents less than three percent of a man-year (54 ÷ 1880), no plan would probably be required to compensate for such an insignificant variance.

For the operations to be fully performed, hours must be worked by someone to make up the variance between the straight-time hours available and the hours required to do all of the work. If the variance were significant (say 10%), plans would be made to meet the situation. An alternative sometimes selected is to periodically check the machine tool oiler operations to see if a backlog is actually developing and then decide what to do about it. This approach delays having to provide a solution until a pressing need actually exists. The main shortcoming of this method is that there may not be any discretionary funding in the budget with which to solve the problem. Assuming that the planning data is valid, the variance is real. To delay in developing a plan to work off the variance merely allows time for a crisis to develop. Thus, to delay the planning is no solution at all and could lead to more problems than it will solve.

A second alternative for handling the variance would be to off-load some of the work to a machine tool mechanic. Although this is sometimes done, it is not the most cost-effective solution. You would be paying a higher wage at a premium to have the work done. This premium is the difference between the lower wage paid to a machine tool oiler and the wage paid a mechanic.

Another alternate plan would be to have the machine tool oilers work overtime. Although this is often done, it is not a cost-effective method. You would be paying a premium price to get the work done. This premium is probably higher than it would be if the work were off-loaded to a machine tool mechanic who was paid his straight-time rate to do the work.

A fourth alternative is to plan to work off the variance through the use of part-time help. This is usually the most cost-effective method. You would be paying the proper wage for the work performed. The use of this method, however, is predicated upon the availability of qualified part-time help and no labor relations restrictions on its use. Thus, this solution is not always tenable even though it is usually the most cost-effective method of working off the variance between hours to be worked and labor hours available to do the work.

The power of the MICRO approach over the MACRO method is that it is considerably more definitive. The major drawback in using MICRO is the fact that a considerable amount of labor must be expended in gathering and computing the detail. Maintenance managers who use the MICRO method will concede that it is time consuming, but they consider it time well spent for three reasons. First, they have definitive information from which to plan their manpower and budgets. Second, they have precise information with which to substantiate their budgets. Third, if the budget is to be slashed, they can specifically identify which work will not get done and can predict what will happen when it does not get done. This capability to specifically present the impact of a proposed budget cut puts those members of management who want to reduce the maintenance department's budget in a position of having to choose which work should not be done. Rather than make this choice, they will sometimes leave the budget alone and make reductions in other departments. Even if the budget is reduced, the maintenance manager would have had an opportunity to educate his superiors on the functions performed by his department. While doing this he can deal in specifics rather than broad generalities. This opportunity would not have been available to him if he had used the MACRO approach in developing his manpower requirements.

USING HOURS OF WORK IN PLANNING

Hours of work is another unit of measure for quantifying plans. It is often useful at the working level in a maintenance department and is readily convertible to a headcount number or dollars. It is a more precise measuring device than headcount, because hours are more easily allocable to specific tasks to be performed than are heads. Allocation by headcount often requires that a person's time be split among various tasks to be performed; the resulting headcount, or man-day, fractions are harder to work with than are hours.

Detailed construction cost estimates are first prepared in terms of man-hours by specific craft skills and materials required. The craft man-hours and quantities of various materials are then converted to dollar values. Cost estimates for other maintenance activities can and should be prepared in the same manner. To merely state that you need eight electricians in the maintenance department is an insufficient presentation. The total required should first be determined in terms of man-hours required for specific operations. These man-hours can then be converted to a headcount number if it is determined that the work is to be done using craftsmen who will be on the payroll. The man-hours can also be converted to dollars if the work is to be contracted.

The MICRO method of headcount planning described in the previous section essentially relies first upon the determination of man-hours required and second upon the conversion of these hours to a headcount number. The use of the man-hour unit of measure is also an effective method of evaluating bids. An unusually low price quotation may indicate that the bidder has failed to comprehend the labor requirements needed to adequately perform the work. Thus, his planning is faulty.

PLANNING PREMIUM-TIME REQUIREMENTS

Planning of in-house labor costs requires consideration of overtime in addition to headcount requirements. Figure 1-7 shows the impact of premium time on costs. As indicated, if the overtime rate is one and one-half times the regular rate and a job is scheduled to work 50 hours per week, the work hours are increased by 25% over a 40-hour week. The payroll cost, however, is increased 37 ½%. Ten percent of the total payroll for hours worked is a premium representing *pay for hours not worked.*

If the rate is one and one-half times the regular rate and a job is scheduled for 60 hours per week, the hours of work are increased 50% over the 40 hours per week, but the payroll cost is increased by 75%. Over 16% of the payroll is premium pay for hours not actually worked. As indicated in Figure 1-7, these increases in payroll costs are even more staggering if a double-time rate is paid for the overtime work.

Cost-effective planning must control overtime as well as headcount. Overtime should normally be applied to those specific situations when an increase in headcount is not warranted, due to a short-range requirement for the additional hours to be worked. Examples of such specific situations are incidents of high absenteeism, inclement weather, peak vacation periods, maintenance crew requirements on paid holidays and weekends, or emergency maintenance.

It is possible to arrive at a dollar figure for premium time by merely multiplying the total overtime hours by the average rate of pay for all craftsmen. This approach, however, lacks precision, first, because there are wage differentials between craft skills and, second, because overtime is not usually equally required among the various crafts.

Assuming that overtime is equally distributed among employees in a craft, the average rate of pay for the particular craft may be multiplied by the total premium hours required to arrive at the dollar figure. As an example of how to make the computations, let us use a plan requiring 3400 hours of premium time for first-shift air conditioning mechanics. If the average base hourly rate for A/C mechanics is $5.30, then a premium rate at time and one-half would be $7.95. Multiplying $7.95 by the

3400 hours yields a premium-time dollar figure of $27,030 for air conditioning mechanics. To arrive at the total budget for the labor of air conditioning mechanics, this premium-time figure must be added to the dollar figure for straight-time pay. In this example, if there are 10 air conditioning craftsmen on the payroll, their straight-time pay for hours actually worked would be $99,640 ($5.30 x 1880 hours x 10 men). Combined with the $27,030 for premium time, this represents a total of $126,670 in direct payroll expense.

PREMIUM PAY IMPACT ON DIRECT PAYROLL EXPENSE

	Premium Pay Rate @ One and One-Half of Base Rate		Premium Pay Rate @ Double Base Rate	
	50 Hours	60 Hours	50 Hours	60 Hours
Increase in Hours Worked	25%	50%	25%	50%
Increase in Direct Payroll Cost	37 1/2%	75%	50%	100%
Portion of Direct Payroll Paid as a Premium	10%	16 2/3%	20%	33 1/3%

FIGURE 1-7

PREMIUM PAY ANALYSIS

The 1880 hours used in the above formula represent the net available hours. Figure 1-8 shows the amounts deducted to arrive at the 1880 hours. In this example, the pay for sick days, vacations and holidays would be $10,600 ($5.30 x 200 x 10 men). These costs are normally treated as related payroll expense along with such items as social security contribution, federal unemployment insurance, etc. If such costs are included in the maintenance department's budget, they must be added to the $126,670 in direct payroll expense for straight-time and premium-time hours of work to arrive at the total budget for in-house air conditioning mechanic labor.

NET AVAILABLE STRAIGHT-TIME HOURS

Gross Straight-Time Hours (40 Hours/Week x 52)		2080
Deductions:		
Vacation (10 Days) =	80	
Sick Leave (5 Days) =	40	
Paid Holidays (10 Days) =	80	200
Net Available Straight-Time Hours		1880

FIGURE 1-8

SAMPLE COMPUTATION OF NET MAN-HOURS

USING OTHER UNITS OF MEASURE

Units of measure relating to the equipment to be maintained can also be used in planning work. Motor vehicle maintenance is usually predicated upon operating miles. For airlines, engine overhaul is based upon hours of operation. Some of the maintenance recommended by manufacturers of heating, ventilating and air conditioning equipment is also based upon hours of operation. Using these units of measure in work planning first requires an identification of what operations will be required at given times. The necessary tools, parts and manpower can then be planned for performance of the required operations.

With the advent of portable or permanent measuring devices on production and building equipment, temperature, pressure and vibration have become units of measure for planning work. These forms of measurement can be used to diagnose a defect while a machine is still running. Planning for tools, replacement parts and manpower can then be effected for a scheduled shutdown. This use of sensing devices is sometimes called "predictive maintenance."

In vibration analysis, as an example, the level of vibration is related to the severity of the defect. At one level the equipment should be shutdown for work immediately, at a lower level within two days, and at still a lower level within ten days. The detection of vibration at the lower level permits the scheduling of a shutdown and prevents a breakdown failure that could damage the equipment. It also eliminates the overtime hours frequently required to perform breakdown maintenance. Vibration analysis is particularly useful on high-speed rotating machinery such as air-handling units.

2 Modern Approaches to the Maintenance Budget—Keystone to Rigorous Financial Controls

Historically, the usual unit of measure for quantifying plans has been the dollar because it is the common denominator for the expression of all activities performed by an organization. In business, dollars on the balance sheet depict the position of the company, a prime concern of top management.

In order to effectively communicate its intentions to top management, a maintenance department should express its plans in terms of dollars. Management is most familiar with this unit of measure and can easily compare departments on this basis.

The most common planning device using dollars is the budget, a cost plan for expenditures over a prescribed fiscal period. A budget is not only the most commonly known plan that uses dollars, it is also the plan most often scrutinized by top management. This chapter describes some of the techniques used in developing budgets, as well as other uses of dollars in planning.

INFORMAL METHODS OF BUDGETING

The most commonly used informal methods of budgeting are:

- Level Budgets
- Current Payroll Plus Estimate of Purchases

The most common reason given for using these methods is their simplicity. Simple solutions to complex challenges, however, are usually proposed by simpletons.

THE DISADVANTAGES OF LEVEL BUDGETS

In essence, a level budget proposes that an organization spend as much in the current year as it did in the previous year. There is no consideration given to obvious external changes in economic factors such as an inflationary spiral.

In an inflationary period, the use of the level budget method amounts to a budget cut and results in a decrease in the level of maintenance department effort. The department has the same amount of dollars to spend, but labor and materials cost more. Thus, the maintenance manager can only get less each year with the dollars he is allocated to spend.

The level budget method also ignores internal factors such as an increase or decrease in the amount of square footage or number of equipment items to be maintained.

Thus, the level budget method lacks realism, since it ignores internal and external factors that affect the mission of the maintenance department. It assumes that what was spent last year is the right amount of expenditure for the coming year.

CURRENT PAYROLL PLUS ESTIMATE OF PURCHASES

The current payroll plus an estimate of purchases method of budgeting is only a little more sophisticated than the level budget approach. The estimate for materials and services portion of this method does allow for some recognition of external and internal factors. The estimate permits consideration of changes in prices and needs for purchased goods and services. However, by holding to the current payroll, it creates a level budget for in-house labor. If wage increases are necessary to maintain the labor force, there must be a decrease in headcount or overtime in order to stay within the dollar limitation. The current payroll portion of the budget also assumes that the workload for the internal maintenance work force will not change. Thus, the current payroll plus an estimate of purchases method of budgeting provides for a recognition of changed conditions in purchasing items, but no consideration of changes in the factors regarding in-house labor.

FLEXIBLE, INFORMAL METHODS OF BUDGETING

The flexible, informal methods currently in use in budgeting for maintenance activities are:

- Fixed Percentage of Plant Investment
- Adjusted Percentage of Plant Investment
- Adjusted Historical Cost

FIXED PERCENTAGE OF PLANT INVESTMENT

The fixed percentage of plant investment technique relates the requirements of maintenance to the dollar value of plant investment. The method assumes that there is a correlation between how much is spent on maintenance and how much capitalized plant investment the organization has. That such a valid correlation exists, however, may be an erroneous assumption dependent upon many factors. For example, changes in an organization's fiscal policy and occupation of a greater or lesser amount of leased facilities would have a marked impact on the percentage. If the organization changes its fiscal policy with respect to leasing rather than buying plant equipment, the percentage would also be affected. If the organization began to change its internal operations (e.g., from providing services to engaging in manufacturing), this factor would also have an impact on the ratio. Also, the fixed percentage method does not consider the obvious factor that the requirements for maintenance and repair expenditures increase as facilities become older.

In addition to the possible erroneous assumption of a valid correlation between maintenance costs and plant investment, the method has the same shortcomings previously discussed with respect to a level budget. The percentage is fixed, thus essentially invoking a level budget if the value of the plant investment does not change. The only flexibility allowed by the method is directly related to changes in the dollar value of the plant investment.

ADJUSTED PERCENTAGE OF PLANT INVESTMENT

Because this type of budgeting permits adjustment factors, it is a better method than the fixed percentage technique. It also avoids the weaknesses of a level budget. The adjusted percentage method, however, although providing for flexibility, still assumes that there is a correlation between required maintenance costs and the amount of plant investment. For a given organization this assumption may not remain valid over a prolonged period of time.

ADJUSTED HISTORICAL COST

Of the flexible, informal methods of budgeting, the adjusted historical cost method is probably the most effective. As with other informal methods, one major advantage is its relative simplicity. The method begins by ascertaining the previous costs incurred for maintenance. This cost figure is then adjusted based upon such factors as:

- Changes in the prices of purchased material and services.
- Changes in requirements for purchased materials and services.
- Changes in direct and indirect costs of in-house maintenance labor.
- Changes in the requirements for in-house maintenance labor.

This flexibility, based upon changing internal and external conditions and requirements, is the primary advantage of the adjusted historical cost method. For smaller maintenance departments it is probably the most cost-effective method of developing a realistic budget. The effectiveness of the method is directly related to the ability to identify what changes have or will occur in the internal and external factors considered in making the adjustments to the historical cost data.

FORMAL METHODS OF BUDGETING

Most formal techniques of budgeting are based largely on the following:

- Type of Item
- Type of Work
- Organization

Any of these formalized approaches requires substantially more time in planning the budget; however, it is time well spent. Planning a realistic budget first requires the planning of all of the activities the budget will support. It gives the maintenance department an opportunity to act rather than react and to identify what is required to carry out its assigned mission. Furthermore, if all the requirements identified within the budget plan are to be executed, the total dollars must be approved. If the proposed budget is reduced, the maintenance manager has the opportunity to specifically identify those requirements that cannot be met, leaving higher management to face the risks involved in not providing the resources for executing them.

In any of the formal methods of budgeting flexibility is assumed. Each year is a new challenge and the costs for that year are planned in the light of what is known with respect to requirements for that year. The past may provide some insight but is not a paramount consideration in the formal planning methods.

USING THE TYPE-OF-ITEM BUDGET

In using the type-of-item method for budgeting, the first step is to establish various cost elements within the total budget. The components selected will vary based upon the functions performed by a given maintenance department and the capability of the accounting system to identify costs by the various elements selected. Figure 2-1 shows an example of a budget using the type-of-item approach. Note in the example that the major cost elements are organized by type of item and that the subcategories for each item are the types of work performed for each item. Regardless of the types of items selected as components, some thought must be devoted to the requirements of accomplishing each element, which must then be quantified in terms of dollars.

THE TYPE-OF-WORK APPROACH

A type-of-work budget is somewhat similar to the type-of-item approach. The difference is that the major cost elements are organized by type of work rather than by type of item. The first step in a type-of-work budget is to establish the various categories that will be used. These elements will vary depending upon the types of work performed by a specific maintenance department and the capability of the accounting system to report costs for each of the various elements selected. Figure 2-2 is an example of a type-of-work budget.

USING THE ORGANIZATION APPROACH

The organization method of budgeting consists of having each supervisory member of the maintenance department prepare his own budget. Thus, the major cost elements within the budget are based upon the organizational structure of the particular maintenance department. The use of this approach requires an accounting system that can identify costs by each section within the maintenance department. A budget is prepared by organizational element using, for example, the three subelements

MAINTENANCE BUDGET ELEMENTS

1. Machine Tools
     Preventive Maintenance
     Emergency Repairs
     Scheduled Repairs
     Rehabilitation/Replacement

2. House Equipment
     Preventive Maintenance
     Emergency Repairs
     Scheduled Repairs
     Rehabilitation/Replacement

3. Buildings
     Maintenance
     Minor Repairs and Services
     Modifications and Rearrangements
     Rehabilitation

4. Land Improvements
     Parking Lots, Roads and Lighting Maintenance
     Landscape Gardening
     Modifications and Rearrangements
     Rehabilitation

5. Supervision and Clerical Support

FIGURE 2-1

SAMPLE TYPE-OF-ITEM BUDGET ELEMENTS

MAINTENANCE BUDGET ELEMENTS

1. Maintenance and Repair
     Machine Tools
     House Equipment
     Office Equipment
     Buildings
     Land Improvements

2. Modifications and Rearrangements
     Buildings
     Land Improvements

3. Rehabilitation/Replacements
     Machine Tools
     House Equipment
     Office Equipment
     Buildings
     Land Improvements

4. Supervision and Clerical Support

FIGURE 2-2

**SAMPLE TYPE-OF-WORK
BUDGET ELEMENTS**

of in-house labor, purchased materials and services, and capital equipment requirements.

One of the main advantages of using the organization method of budgeting is that it requires the participation of each section supervisor within the maintenance department. Each supervisor must identify what he is going to do and then quantify these efforts in terms of dollars. Figure 2-3 is an example of the budget elements of a maintenance department that is essentially organized by craft skill.

USING DOLLARS FOR OTHER PLANNING

The use of dollars in planning is not restricted to budgeting. For example, the cost estimates prepared for each alternate solution to a given maintenance engineering problem in essence represent a forecast of the costs that would be incurred in the

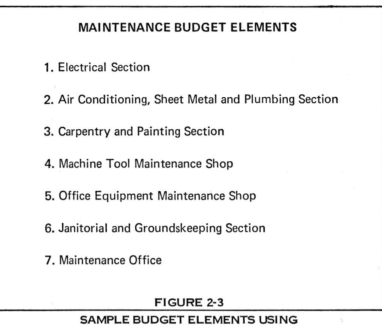

FIGURE 2-3

SAMPLE BUDGET ELEMENTS USING
THE ORGANIZATION APPROACH

execution of each proposed solution. Obviously, not all solutions will be selected. All other things being equal, the one predicting the least costs will be selected.

Dollars can also be used effectively in comparing the cost-effectiveness of various levels of a maintenance effort. Figure 2-4 shows the costs of different frequencies of window washing at an airport terminal building. Note that in using this method a base cost is first established for the activity, shown at the top of Figure 2-4. This base cost is extended for the various frequencies of the activity in the bottom portion of the figure. Note also that the cost is subtotaled into the separate categories of "landside" and "airside" to facilitate rapid calculations of the costs of selecting different frequencies for each subcategory. In this airport example, the "airside" glass generally will become dirtier more quickly than "landside" glass because the exhaust from jet aircraft will film up "airside" glass, but does not usually affect the "landside" glass. This factor should be considered in establishing the level of maintenance for window washing. One alternative would be to wash the "airside" glass more often than the "landside" windows.

In an effort to maximize short-range profits, maintenance activities are sometimes cut severely, resulting in an actual budget that is considerably less than the expenditure necessary to properly maintain the facilities. One of the best ways to present the monetary needs of the maintenance department is to forecast the culminating effect of improper budgeting in terms of future cost to the organization. These costs can be expressed as dollars related to lost production man-hours resulting from equipment breakdowns; the decline in the useful lives of equipment, which will require earlier capital replacement or rehabilitation due to improper maintenance; the added capital investment in landscaping required to replace items lost by a niggardly groundskeeping program; or any other factor pertinent to the mission of the maintenance department. Thus, dollars can be used to tell an organization what it should spend as well as what it will spend for maintenance.

EXTERIOR WINDOW WASHING—MAIN AIRPORT TERMINAL BUILDING

Cost of One Complete Wash

Landside	Airside	Total
$180	$216	$396

Annual Cost Based on Frequency

Frequency	No. of Times	Landside Cost	Airside Cost	Total
Every 2 Weeks	26	$4,680	$5,616	$10,296
Every 3 Weeks	17	3,060	3,672	6,732
Every 4 Weeks	13	2,340	2,808	5,148
Every 2 Months	6	1,080	1,296	2,376
Quarterly	4	720	864	1,584
Semiannually	2	360	432	792

FIGURE 2-4

SAMPLE USING DOLLARS TO COMPARE LEVELS
OF MAINTENANCE EFFORT

3 *Planning the Level of Maintenance Effort*

The term "level of maintenance effort" refers to what a maintenance department should be doing to achieve its mission. The health and safety of the public and employees, minimization of downtime hours on production or building equipment, housekeeping and keeping the general appearance of the facilities presentable are normally all part of the assigned responsibilities of the maintenance department. The ultimate goal should be to optimize the use of every maintenance dollar spent in achieving the assigned mission.

The most effective means of planning the level of maintenance effort will vary depending on the size and mission of a given maintenance department. There is one certainty, however: if only rudimentary methods are applied, there will be little opportunity for achieving a cost-effective maintenance program over the long run. This chapter tells you the reasons for this certainty. It also describes methods of planning the level of maintenance effort to achieve cost savings, permitting a maintenance department to plan what it wants to do, determine what it is doing and reconcile the differences between the two by additional planning.

RUDIMENTARY APPROACHES

The rudimentary approaches to maintenance effort virtually amount to unplanned systems. The three levels of sophistication are:

- User (do-it-yourself)
- Breakdown (time of occurrence)
- Random (haphazard assignment)

USER DOES IT

Under the user approach, whoever is using a piece of equipment within an organization performs the maintenance. If he can't fix it, he then asks the maintenance department for help. Obviously, if the user is not trained in maintenance for the piece of equipment, his efforts may do more harm than good. This constitutes the major drawback to this approach. Further, the user may be in higher job classification than the maintenance craftsmen who would do the work, which means the organization is paying more for the maintenance. If the user is on an incentive-type pay plan, there is even less chance that any maintenance will be performed. The user is concerned with getting out production units and will not be willing to spend any time on maintenance.

BREAKDOWN MAINTENANCE

The breakdown approach to maintenance in essence says you don't do anything until a piece of equipment fails. At the time of a failure, a maintenance man then responds to a call for service. There are some maintenance departments that still operate strictly on a breakdown basis. Their work is unplanned and they merely react to calls for service. The main disadvantage to this approach is that it is more costly over the long run. The attendant repair and rehabilitation costs for damage to equipment under a breakdown program are always more costly. Depending upon the type of equipment, these costs can be as high as 300% over an extended time period.

There are still some maintenance managers who say, "When my men have nothing to do, the facilities are running well and we are doing a good job." This statement is a holdover from the days of the typical, mismanaged maintenance department where craftsmen sat around and remained "on call" until an emergency arose. Certainly under this approach repairmen get onto a job very quickly to get the equipment operating again, but at a tremendous cost in standby labor time. In planning the level of maintenance effort, one target should be to eliminate standby labor. This can be done by assigning these mechanics, on a formal and documented basis, to such tasks as scheduled inspection and servicing of building or production equipment, overhaul or repair of spare parts or equipment in the shop area, or servicing and overhaul of tool crib equipment. These types of jobs can be dropped when a true emergency occurs and can be picked up later without costly delays.

RANDOM MAINTENANCE

The haphazard or random approach usually takes the form of having enough manpower available in the maintenance department so that, on occasion, not everyone is responding to emergency calls. In these slack moments a craftsman may be assigned to "take a look" at some equipment. Since these slack moments occur on a random basis, the maintenance on those pieces of equipment that are checked is haphazard at best. Since the maintenance man is not usually told specifically what to do, he may not do anything. He may come back and tell his foreman that something "sounds bad." In turn, the foreman may forget this verbal report or, due to more pressing matters, fail to have anything done before the item breaks down. Thus, a random approach means that occasionally some maintenance is pulled, but there is no precision in the process. The craftsman is not accountable for any work and there is usually no record of what was done.

Random maintenance is sometimes referred to as "sleep on the roof" maintenance. It got this name from an incident that occurred at a plant in Virginia. During the typical, drowsy summer afternoons when things were slack, the maintenance mechanics would just as soon go fishin'. However, since they wanted to be paid and had to be at the plant in case of an emergency call, they were assigned to "inspect and service roof equipment." For the mechanics, this assignment developed into the task of going up on a roof and taking a nap. It was not until a new shift foreman decided that the assignments should be formally scheduled using work orders and checked out the activity that the siesta time was abruptly eliminated.

ELEMENTARY APPROACHES

Under elementary approaches some attempt is made to describe the level of maintenance effort in terms of what types of work are to be done and when they are to be done. The various techniques that may be applied are:

- Shift Assignment
- Repetitive Operations Identification
- Work-in-Progress Files
- Backlog Control

APPLYING SHIFT ASSIGNMENT

One elementary form of planning identifies certain types of work by specific shifts. Examples of this approach are as follows:

- Office equipment service is designated as a first-shift operation to facilitate coordination between the repairman and the equipment user.
- Janitorial floor refinishing is performed on the third shift so that the areas are not in use while the work is being done.
- Interior painting is scheduled for the second shift when the areas to be painted are not occupied.

IDENTIFYING REPETITIVE OPERATIONS

After designating the shift, a more sophisticated planning system for repetitive operations will identify specifically where the work is to be done and the prescribed frequency for the effort. For example, janitorial floor refinishing of public corridors on the first floor of a given building is to be done every four weeks. When preventive maintenance on equipment is included, certain work is designated to be done at prescribed intervals. Normally, this maintenance covers periodic lubrication, inspection of moving parts and replacing of items such as fan belts and filters.

USING A WORK-IN-PROGRESS FILE

The purpose of a work-in-progress file is to provide visibility over the current level of effort within a maintenance department, that is, to show who is working on what. The system should be planned to identify who, what, and where work is being done. One of the most effective means of accomplishing this system is to use work authorizations.

One maintenance department instituted a multi-copy work order form that could be used for scheduled maintenance requirements, trouble calls on equipment failures or authorization for repairs. For whatever purpose the work order was issued, the foreman kept a copy. These copies were filed chronologically in a separate file for each craftsman assigned work and the duplicate copies were given to the craftsmen. When a craftsman completed a job he returned the work order to the foreman who in turn, removed his duplicate copy from the file. With the file the foreman could quickly ascertain the status of all work that had been assigned to each craftsman for which completed orders had not been returned. Follow-up could also be accomplished to determine the status of the work.

In a larger maintenance organization a similar system was used with two files. One file maintained by the foreman had copies of the work orders filed by building, and the other file maintained by the leadman had work order copies filed by the craftsman assigned the work. Under this system the leadman tracked the work of the individual craftsman while the foreman tracked all work by building.

In either of these examples, existing paperwork is used for a work-in-progress file rather than introducing more forms into the manual system.

APPLYING BACKLOG CONTROL

Backlog reports are a very effective tool for planning both on a long-term and on a day-to-day basis. Despite this, many maintenance departments today do not have adequate backlog records available, thus supervisors have very little factual information about the workload and the current backlog of jobs yet to be done.

Some form of workload record is vital to planning, even if it consists only of a count of the number of work orders waiting to be started. Much more useful, however, is a current backlog report showing the number of standard hours covered by work orders that are to be completed by the maintenance department. How much backlog is acceptable depends on the operation. For one department, a three- to four-week backlog may be acceptable; for another, an eight- to ten-week backlog may be appropriate.

The backlog report is a useful planning tool for two reasons: first, it can be used to identify a craft that may become a bottleneck in project completion; second, it can identify trends indicating the need for management action.

Figure 3-1 is an example of a backlog report that identifies the backlog of hours by craft skill. These hours of work can be compared with the available hours for each craft to determine where a substantive backlog is developing in any given area. For example, note that in Figure 3-1 the backlog in sheet metal work for the week ending 24 January is 1694 hours. If the sheet metal craft employs only four men, normally only 160 hours of straight time are available per week. This would mean that the sheet metal craft has more than a ten-week backlog ($1694 \div 160 = 10.58$). In comparison, the carpenters have 1723 hours backlogged. If there are ten carpenters, this means there are 400 hours normally available per week, which results in less than a five-week backlog ($1723 \div 400 = 4.3$). If the sheet metal work is an intricate part of modification

WEEKLY BACKLOG REPORT

YEAR: 19XX

| WEEK ENDING (FRIDAYS) | ELECTRICIANS | SHEET METAL MECHANICS | CRAFT HOURS | | CARPENTERS | PAINTERS | TOTAL |
			PLUMBERS	AIR CONDITIONING MECHANICS			
3 January	1696	1210	3015	1945	1707	1301	10,874
10 January	1703	1365	3017	1947	1710	1298	11,040
17 January	1709	1575	3027	1945	1715	1295	11,266
24 January	1715	1694	3039	1936	1723	1303	11,410

FIGURE 3-1

SAMPLE BACKLOG REPORT

jobs that involve the other crafts, these figures mean that these projects will be paced by the ability of the sheet metal mechanics to get their part of the work done.

Once identified, there are several alternatives available to alleviate a bottleneck. Some of these solutions are the use of overtime, contracting some of the sheet metal work, bringing in purchased labor, or giving first priority to the sheet metal work which is pacing the other crafts on projects.

Backlog reports designed to indicate trends are also valuable in identifying the need for additional planning. If the backlog continues to increase, action should be taken. Some of the work may have to be deferred, other jobs may have to be contracted or the hiring of additional craftsmen can be effected based upon the factual justification provided by the backlog report.

On the other hand, if the backlog continues to decline from period to period, action should also be taken. If the backlog drops too low, craftsmen will be idled by lack of work. To prevent this from occurring, it may be desirable to start on otherwise deferred maintenance projects, bring in work that is currently contracted or eventually reduce the number of maintenance men on the payroll.

Figure 3-1 displays numerical values on a weekly time increment which can be analyzed to determine short-term trends. For example, examination of the backlog for sheet metal mechanics reveals that there has been an increasing rise in the backlog over the short term. From the week ending 3 January to the week ending 24 January, the backlog has increased from 1210 to 1694 hours. The rate of growth has been in excess of 120 hours per week.

A more effective way of presenting this short-term trend data would be to use a chart or graph. Figure 3-2 is an example of a short-term graph displaying the same numerical values shown in Figure 3-1. Note how much easier it is to determine what has been happening to the backlog on a short-term basis by using a graphic representation.

For long-term trend data, the use of a chart or graph is definitely a better means of displaying the information than a written report. Figure 3-3 is an example of a long-term trend chart. Note that the time increment used is months whereas Figures 3-1 and 3-2 used a time increment of weeks. For long-term trend analysis the use of a monthly time increment smooths out some of the sporadic peaks that may occur on a weekly time increment chart due to holidays, sick leave or vacations and provides a more clear-cut display of what is happening over the long run. Also note that in Figure 3-3 the monthly backlog value for January in the third year is an average of the backlog hours that appeared in Figure 3-1 for the four weeks ending in January. That is, it is the average of the values 1210, 1365, 1575 and 1694, which equals 1461 hours $(5844 \div 4)$.

The value of long-term trend analysis is revealed by an examination of Figure 3-3. The chart shows that since January of the first year there has been a steady, increasing rise in the backlog in the sheet metal craft. Thus, the existing backlog in excess of ten weeks is not an unusual occurrence that can be readily resolved by having the four sheet metal mechanics merely working overtime for a while. Either an additional man should be hired or plans made to contract some of the work on a regular basis.

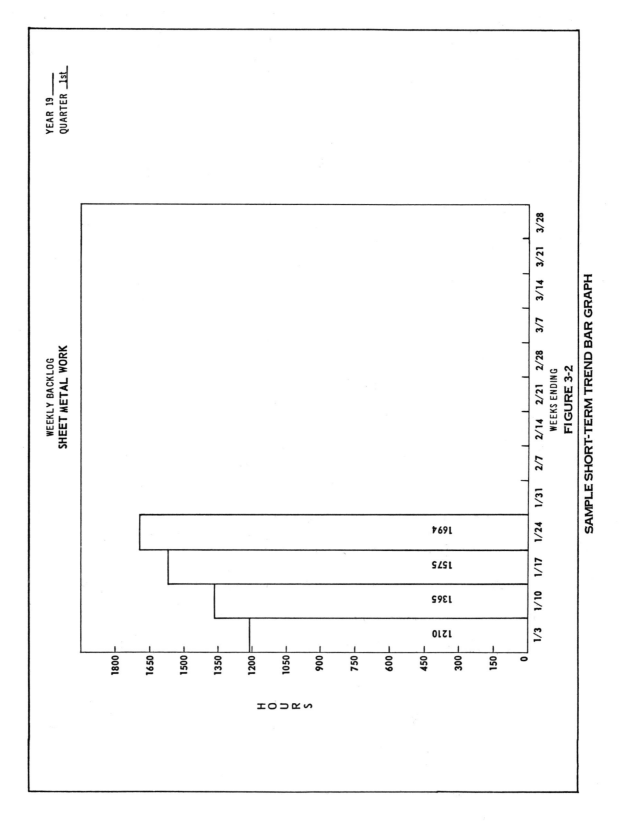

SAMPLE SHORT-TERM TREND BAR GRAPH

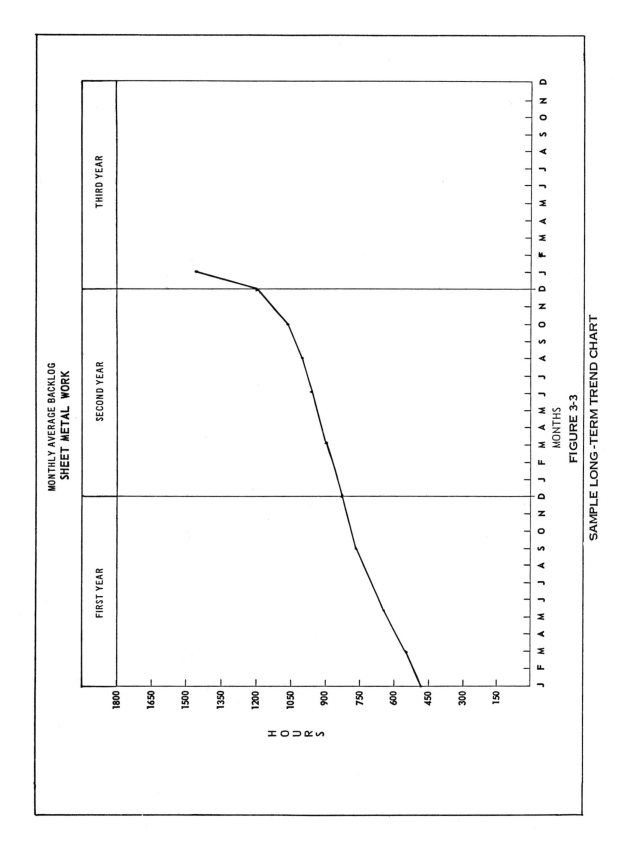

FIGURE 3-3

SAMPLE LONG-TERM TREND CHART

51

COMPREHENSIVE PROGRAMS

Under comprehensive programs for planning the level of maintenance effort, all work is examined for the allocation of resources. Resources are allocated to all requirements for the types of activities performed by the maintenance department and a feedback reporting system is used to compare work actually performed with the planned levels of effort.

Methods used for a comprehensive program will vary based upon the size and mission of an individual maintenance department. Some of the successful methods of achieving a comprehensive maintenance program are:

- Cost Allocation
- Labor Hours
- Equivalent Heads
- Integration of Operations

USING A COST-ALLOCATION SYSTEM

Cost-allocation planning of the level of maintenance effort can be a highly effective approach. It is particularly useful when a substantial portion of the maintenance effort is contracted and dollars are the best unit of measure by which to plan and then compare actual performance. An effective system requires a breakdown of total maintenance costs into meaningful components and an adequate reporting method of actual expenditures.

The amount of detail desirable and the best manner of separating the components in the breakdown will vary depending upon the needs of the particular maintenance department. Figure 3-4 is an example of a cost-allocation breakdown that uses type-of-work components. The three classifications are emergency maintenance, scheduled service and inspection, and scheduled repairs. Each of these three types of work is expressed as a percentage of the total.

The cost-allocation report can serve several purposes: first, it is useful in identifying trends; second, it can determine if events are conforming to plans and where adjustments in plans may be desirable.

For example, if a scheduled inspection and maintenance program on machinery and equipment is being effective, in the long run there should be an increase in the number of scheduled repairs and a decrease in more costly emergency repairs. If the planned goal is 20% for emergency maintenance, 30% for scheduled inspection and maintenance, and 50% for scheduled repairs, the actual dollars expended for each of these components can be tracked both on a trend and a comparative basis using a breakdown report. As indicated in Figure 3-4, the performance during the month of March is closer to the goal than the total year-to-date performance. If performance in subsequent months comes closer to the plan, the year-to-date cumulative performance would also reflect this change.

COST OF PLANT AND ACTUAL EQUIPMENT MAINTENANCE

MONTH OF: MARCH 19XX

TYPE OF WORK	ANNUAL PLAN DOLLARS	%	EXPENDITURES THIS MONTH DOLLARS	%	EXPENDITURES YEAR-TO-DATE DOLLARS	%
Emergency Maintenance	120,000	20	12,650	23	37,250	25
Scheduled Service and Inspection	180,000	30	15,950	29	44,700	30
Scheduled Repairs	300,000	50	26,400	48	67,050	45
TOTALS	600,000	100	55,000	100	149,000	100

FIGURE 3-4

SAMPLE OF COST-ALLOCATION SYSTEM REPORT

LABOR HOURS REPORT

WEEK ENDING: 4 July 19XX

CRAFTS

Type of Work	Electricians	Sheet Metal Mechanics	Plumbers	Air Conditioning Mechanics	Carpenters	Painters	Total	%
Emergency Maintenance	28	12	20	80	8	0	148	7
Scheduled Service and Inspection	32	0	8	240	20	0	300	13
Scheduled Repairs	60	16	40	80	60	160	416	19
Construction	200	128	60	176	200	160	924	41
Vacation, Sick Leave, Paid Holidays, etc.	80	24	32	144	80	80	440	20
Total	400	180	160	720	368	400	2228	100

FIGURE 3-5

SAMPLE LABOR-HOURS BREAKDOWN

APPLYING A LABOR-HOURS BREAKDOWN

A labor-hours allocation system is similar to the cost-allocation approach in that the reporting is a breakdown into selected components. However, the unit of measure used in the breakdown is labor hours rather than dollars. The labor-hours approach is particularly valuable where most of the maintenance work is performed by in-house labor. If the level of maintenance effort is largely a function of the allocation and utilization of this labor, the labor-hours breakdown can be more precise than using dollars as the unit of measure.

Figure 3-5 is an example of a labor-hours breakdown using type of work categories as in Figure 3-4 with an additional category for hours charged to vacation, sick leave, paid holidays, and other nonwork days. This category may be used to account for all hours paid and to reconcile the report to payroll records. Note that since the sample report covers the week ending 4 July, the paid holiday is reflected in this additional category. The labor-hours breakdown may be used for both trend analysis and for month-to-month identification of deviations from planned allocations to the various components.

USING EQUIVALENT-HEADS ANALYSIS

An equivalent-heads breakdown is a variation of the labor-hours method. The unit of measure is headcount derived from labor hours. For example, an equivalent-heads value for 180 hours of labor in a given week is 4.5 if a standard 40-hour week is applied ($180 \div 40 = 4.5$). Figure 3-6 is an example of an equivalent-heads analysis using the labor-hours data that appeared in Figure 3-5.

The equivalent-heads approach is particularly valuable if the bulk of the work utilizes in-house labor and a substantial amount of premium time is being expended. When the hours of straight and premium time are expressed as headcount, the opportunities for hiring additional personnel and accomplishing the work at less cost becomes more readily apparent. For example, in Figure 3-6 note that total sheet metal effort used an equivalent-heads value of 4.5. If there are only three sheet metal workers, this means that 60 hours of premium pay were expended to get the additional labor hours equal to 1.5 heads. If this were the case over a protracted period, it would probably be more appropriate to hire one additional sheet metal worker or contract some of the work rather than to continue to incur the premium pay costs to get these labor hours of sheet metal work.

INTEGRATING MAINTENANCE WORK

Another facet of comprehensive planning of level of maintenance effort is the integration of maintenance activities with other business operations. A maintenance department cannot operate in a vacuum and expect to receive much appreciation from the other departments of an organization.

EQUIVALENT-HEADS REPORT

WEEK ENDING: 4 July 19XX

Crafts

Type of Work	Electricians	Sheet Metal Mechanics	Plumbers	Air Conditioning Mechanics	Carpenters	Painters	Totals
Emergency Maintenance	.7	.3	.5	2.0	.2	0	3.7
Scheduled Service and Inspection	.8	0	.2	6.0	.5	0	7.5
Scheduled Repairs	1.5	.4	1.0	2.0	1.5	4.0	10.4
Construction	5.0	3.2	1.5	4.4	5.0	4.0	23.1
Vacation, Sick Leave, Paid Holidays, etc.	2.0	.6	.8	3.6	2.0	2.0	11.0
Total Equivalent Heads	10.0	4.5	4.0	18.0	9.2	10.0	55.7
Planned Equivalent Heads	10.0	3.6	4.0	18.0	9.0	10.0	54.6
Variance Actual to Plan	0	+.9	0	0	+.2	0	+1.1
Actual Headcount	10	3	4	18	9	10	54.0

FIGURE 3-6

SAMPLE EQUIVALENT-HEADS BREAKDOWN

Lack of integration was shown in the example of the maintenance manager of a manufacturing company who arbitrarily scheduled a shutdown of the administration building to do electrical and house equipment maintenance work. It so happened that the weekend of the shutdown occurred at the same time as the accounting department's year-end closing. When the accounting department personnel showed up for work on Saturday morning, they had no lighting or electrical power for the calculating equipment. Needless to say the comptroller was less than enchanted with the situation and service was restored to the building. However, man-hours were lost in both the accounting and maintenance departments as a result of this lack of coordination.

Building shutdowns should be coordinated with the users of the facilities so that they can plan their work around the shutdown. Otherwise, the maintenance manager will be confronted by some very unhappy people. No preventive maintenance program on production equipment has ever been completely successful until the production foremen have cooperated in releasing the equipment for maintenance. Likewise, a janitorial floor refinishing or an interior painting program cannot be effectively done unless the job can be scheduled for a time when the space is not being occupied.

Any of these activities requires coordination with the people who use the facilities. All maintenance managers are confronted with the challenge of achieving such coordination. Those managers who have had success have found that the most effective way is simply to *put the plan in writing.*

Whenever the work impinges on the operations of other departments, the plan should be coordinated for concurrence, then stated in written form and distributed to all concerned members of management. To assure widespread notification, one maintenance department puts a notice for a scheduled building shutdown on the bulletin boards and entrances of the building a week before shutdown is to occur. When site roads or parking lots are to be closed for work, the schedule is published in advance in the weekly company newspaper as well as displayed on bulletin boards and entrances. Although some people may not be happy about the necessity for such occurrences, at least personnel are informed that they are to take place and can plan for them.

Figure 3-7 is an example of a notice regarding the temporary closure of a parking lot for resurfacing. Note that the closure involved only one weekday with a Saturday and Sunday used for the curing process after the resurfacing was completed, which would have less impact on parking than if the job had been done on three weekdays.

NOTICE

TO: ALL EMPLOYEES

THE EAST PARKING LOT WILL BE CLOSED FOR RESURFACING ON FRIDAY, SATURDAY AND SUNDAY (JUNE 27TH, 28TH, AND 29TH, 19XX).

PLEASE USE THE WEST PARKING LOT AS AN ALTERNATE PARKING AREA DURING THESE THREE DAYS.

THE MAINTENANCE DEPARTMENT REGRETS ANY INCONVENIENCE CAUSED BY THIS ACTIVITY. YOUR COOPERATION WILL BE APPRECIATED.

FIGURE 3-7

SAMPLE NOTICE OF MAINTENANCE WORK

4 *Proven Methods for Identifying Maintenance Work to Be Done*

The term "identification of work" pertains to how a maintenance department determines requirements for craft labor hours. This chapter presents the various levels of sophistication that can be used to identify work. There is no one best method. Each maintenance manager should select the methods that are best for his organization predicated upon its size and the scope of its mission.

The various methods of identifying work are:

Complaints	Inspection by Craftsmen
Emergency Service Calls	An Inspection Group
Foreman's Inspections	Formalized Planning on Major Jobs
Manufacturer and Safety Standards	Formalized Planning on Minor Jobs
Periodic Maintenance	Work Identification Numbering

COMPLAINTS

The simplest method of identifying work is by getting complaints from the people who are using the facilities. If this is the only method used to identify work, the maintenance department will be in a constant reaction mode. Work is performed willy-nilly with no labor-hour planning other than that necessary to solve the immediate crisis.

EMERGENCY SERVICE CALLS

Although every maintenance department can expect to get a certain number of calls for emergency service, relying solely upon this approach to identify work required is virtually the same as using the complaints method. The maintenance department is geared to a reaction mode of doing business.

FOREMAN'S INSPECTIONS

This technique of identifying work relies upon supervisors to determine requirements for work based upon inspections. The system lacks precision because it assumes that a foreman has the time to walk the facilities and can identify labor-hour requirements from a cursory visual inspection. For some types of work such as cosmetic painting or minor carpentry repairs this may prove an effective technique through the application of simple standards. However, for inspections where machinery must be shut down or disassembled in order to perform the inspection, the use of this method is not appropriate.

USING MANUFACTURER AND SAFETY STANDARDS

For certain types of equipment, the safety required by local ordinances dictates work identification. Elevators must receive inspection in order to obtain annual operating permits. Cranes must be periodically weight-tested to be certified for use. Even if this type of work is contracted, the requirements for such work must be identified before the contracts can be let.

Manufacturer's standards are applied to the maintenance of new equipment items in order to protect warranties. Using this technique, a maintenance schedule that fulfills the manufacturer's recommendations is established for the piece of equipment. In turn, the work to be done dictates the craft skills required. When standards are then applied, the work is identified in terms of craft labor hours required.

APPLYING A PERIODIC MAINTENANCE SYSTEM

When labor standards are applied to a periodic maintenance system, the craft hours required to do the work become identifiable. This identification is achieved by multiplying the labor standard by the number of times the scheduled operation is to be done by the specified craft skill during a given period time. The total man-hours required for each craft skill are then determined by adding up all the scheduled operations to be performed during the time period.

As a simplified example, let us assume that a machine oiler takes two-tenths of an hour to perform a weekly machine lubrication on a given machine. This means this scheduled operation requires 10.4 hours per year (.2 hour x 52 weeks = 10.4 hours). If there are 200 machines that require this same operation, then 2080 man-hours of machine oiler time would be required annually (10.4 hours x 200 = 2080 hours).

PERIODIC INSPECTIONS BY CRAFTSMEN

A well-developed periodic maintenance system should not only require performance of certain maintenance work, but also inspection accompanied by written feedback from the craftsmen describing other work that may be required. This written inspection report may then be used to get an overview of additional work required and to identify the craft skill hours necessary to do the work.

In some instances a separate form is used to provide this written feedback on inspections. However, since most craftsmen do not like to do a lot of paperwork, the technique of providing space for feedback on the work order used for the periodic maintenance work accomplished during the inspection facilitates the reporting process. It also eliminates the need for another form.

Inspection reports should be in writing. Verbal reports are too often forgotten, which negates the effort spent in performing the inspection in the first place.

HAVING AN INSPECTION GROUP

In some instances a separate group of inspectors is established to identify required work. Having the inspectors in a separate section assures that personnel in the group will not be assigned to other tasks. This technique is usually applied by organizations engaged in operations that have a critical need for a high level of health and safety or for which equipment breakdowns have a tremendous impact on profits.

An example of a well-developed inspection section is the one created by a large amusement park in Southern California that was experiencing a high frequency of equipment breakdown. If a ride was not working, it resulted in a direct loss of revenues. More important, however, was concern for public safety. To alleviate the situation, a separate inspection group was established. The primary function of these craftsmen was to inspect equipment for safety and operability. Specific inspection schedules were planned for all pieces of equipment and inspection routes were established to reduce travel time. The inspector's job involved identifying and reporting where maintenance work was required. The only time the inspector actually did any work was in matters of safety where minor adjustments could be performed to remedy the situation. Otherwise, the inspector's reports were issued as work orders to maintenance craftsmen. Performance of the required work was controlled by a feedback of completed work orders. The inspection group proved to be a successful solution; downtime on rides was reduced about thirty-five percent and the frequency of critical equipment failures was decreased by fifty percent.

USING FORMALIZED PLANNING ON MAJOR JOBS

Where the work to be performed by the maintenance department involves substantial effort such as facilities modification or rehabilitation, formalized planning and estimating techniques should be used. This would include preparation and review of plans and specifications and then identification of the number of man-hours required by various crafts to do the work. The personnel designated to identify these requirements for craft labor hours depends, of course, on how the maintenance department is organized. Regardless of who performs the function, however, this system provides for the identification of work on major jobs.

USING FORMALIZED PLANNING ON MINOR JOBS

There are those who advocate the use of the formal planning and estimating technique to identify work on smaller jobs. These methods generally rely upon the use of job planners. They walk the jobs and perform a detailed analysis to determine the craft hours required. The use of job planners on minor work is predicated upon the fact that often half the jobs done by a maintenance department entail five hours or less. Thus, to achieve proper identification of work and control of operations, it is deemed appropriate to perform detailed planning and estimating on these smaller jobs as well as on major ones.

Consulting firms generally advocate this type of approach. They want to sell to maintenance departments their services of establishing such techniques. Thus, they must convince potential clients of the need for their services by advising the use of job planners on minor work, which is a sizable part of maintenance department responsibility. This does not mean that the technique is ineffective. In some instances it has resulted in increases in productivity of twenty to twenty-five percent. However, it is not necessarily a technique of identifying work that should be adopted by all maintenance departments.

The judgment to use job planners on minor work should be based upon the fact that gains in productivity will offset the labor costs of the job planners. This is normally only achievable in maintenance departments employing 200 craftsmen or more. The ratio of minor job planners to craftsmen usually averages 1 to 25. This ratio may vary depending upon the craft. For example, for painters fewer planners may be required and the ratio may be 1 to 35. By the same token, for maintenance electricians the ratio may be 1 to 15.

Those maintenance managers who do not advocate the use of planners for minor jobs generally are opposed to the technique for two reasons. First, they feel that they are already paying for this capability in qualified journeymen and leadmen who will be doing the work. Second, the job planner's work methods may not coincide with how craftsmen feel the job should be done. This can result in a marked variance between the hours of work identified by the job planner and the actual hours spent on the job, which negates the value of a planner to identify work on minor jobs.

APPLYING WORK IDENTIFICATION NUMBERS

The purpose of work identification numbering is to provide a classification system for the various types of work done by a maintenance department. The system is used to accumulate actual labor hours spent on each of the various categories by each craft skill. This historical data is then used to project future craft labor requirements.

MAINTENANCE WORK IDENTIFICATION CODES

Site

10	Grounds
11	Parking Lots and Driveways
12	Lawn Sprinkler Systems
13	Parking Lot Lights
14	Interior Planters and Plants
15	Painting—Curbs and Roadway Stripes

Buildings—Structural

20	Roofs
21	Painting—Inside
22	Painting—Outside
23	Floors
24	Walls
25	Windows and Blinds
26	Doors
27	Ceilings

Buildings—Electrical

30	Lighting System
31	Power System
32	Telephone, Alarms and Communication Systems
33	Instrumentation Cabling and Duct

Buildings—Mechanical

40	Refrigeration System
41	Heating System
42	Air-handling System
43	Plumbing System—Including Sewers and Fire Sprinklers
44	Fuel Storage and Distribution
45	Steam Storage and Distribution
46	Sheet Metal Work

Building Services

50	Janitorial Services
51	Fire and Safety Devices
52	Material and Equipment Handling

Supply Operations

60	Storeroom Operations
61	Receiving Operations

Production Machinery and Equipment

70	Scheduled Lube and Check
71	Mechanical Repairs
72	Electrical Repairs
73	Installations, Mechanical and Electrical Equipment
74	Inspection and Calibration
75	Painting
76	Furniture Repair Service Call

Special Services

80	Requests for installing or relocating blackboards, bookcases, etc.
81	Moves

FIGURE 4-1

SAMPLE WORK IDENTIFICATION CODES

Figure 4-1 is an example of a work identification numbering system used by a maintenance department for a multi-building site in California. The application of the amount of numbers used in this classification system is facilitated by a computerized work order system. In a manual system, extensive clerical labor would be required to maintain such a volume of data, so the use of many numbers would probably not be warranted. The value of the visibility gained by using a large array of numbers should be weighed against the costs of maintaining the data in determining how many classification numbers to use.

As with any forecast based upon past events, the technique of work identification numbers operates on the assumption that future events will coincide with past events. Such an assumption may lack validity. To assure validity in the projection, the historical data should be analyzed and adjusted where appropriate to better reflect future requirements. For example, in Figure 4-1 under "Buildings-Structural" an identification number 20 is given for roofs. The roof on a given building may have been old and a lot of maintenance performed on it during the previous year. However, in the last quarter of the year a new roof had been installed. This would eliminate future maintenance on the roof in the coming year. The historical labor hours for the previous year's efforts should not have to be repeated and should not be included in the forecast of labor hours for roof maintenance.

5 Cost-Saving Work Scheduling and Classification Systems

A scheduling system is a planning tool. Its purpose is to assign resources to accomplish work. What kind of scheduling system should be used by a given maintenance department depends upon the scope and nature of the department's mission. There is no one best way. This chapter presents various methods available to meet the challenge of matching resources with work requirements.

USING FIRST-IN-FIRST-OUT (FIFO)

The first-in-first-out scheduling method processes all work orders sequentially as they are received. This means that all requests for work are treated as having an equal priority. Realistically the FIFO approach cannot be applied to all maintenance operations because all work should not have equal priority. For example, a request to repair a broken hot water pipe would naturally take precedence over a request to replace a washer in a dripping faucet. To some degree, the FIFO method of scheduling can be applied to work that is not critical to operations such as cosmetic interior painting of offices or servicing of office equipment. Even in these examples, requests from executive areas are generally placed ahead of other orders due to political expediency. Thus, FIFO can seldom be completely practiced because strict adherence to the method lacks realism. All maintenance work cannot be assigned equal priority if operations are to be performed efficiently.

APPLYING CONVENIENCE ASSIGNMENT SCHEDULES

The purpose of the convenience assignment method of scheduling work is to reduce operating costs connected with travel, site or fabrication setup costs or design costs. Examples of this method are:

- Grouping work requests to be performed in the same geographical area.
- Combining a fabrication or painting job with a similar job to be performed in a maintenance shop.

When work orders are grouped by geographical area, the travel time of repetitive trips to the same area is eliminated, since all the work pertinent to several work orders is accomplished on the one trip. Examples are requests for minor services such as installing bulletin boards, butt kits and pencil sharpeners. Site setup time may also be reduced for some maintenance operations such as painting. When requests for work in the same geographical area are combined into one work order, design costs may be substantially reduced as well as travel time and site setup time. One department estimated a savings in design and craft labor in excess of $25,000 annually by using this technique.

Combining work orders to reduce fabrication setup costs is usually accomplished in cabinet-making or spray paint activities.

USING A FIXED PRIORITY BY WORK CLASSIFICATION

Use of this scheduling method involves establishing an order of precedence for the accomplishment of work based on a work classification system. An example of such a classification system follows:

First Priority—Emergency/Breakdown Work
Second Priority—Modification Work
Third Priority—Preventive Maintenance

Using the above example, emergency/breakdown work would be defined as work that must be immediately accomplished to prevent damage to facilities or personnel or to restore service to equipment items that failed. This type of work would take precedence over all other work. The next work classification would be modifications. Assuming that all calls for emergency service were being handled, available manpower would then be assigned to modification work. Finally, any additional manpower available would be assigned to preventive maintenance.

One of the requirements of the fixed priority by work classification method of scheduling is that all men in the maintenance department possess an approximately equal capability of performing all work assigned. Each man is a jack-of-all-trades. In smaller maintenance organizations where this is sometimes the case, this method of scheduling is usually found.

When the facilities being serviced require extensive training and knowledge, specialization by craft precludes extensive use of the fixed-priority method. The men in the maintenance crew can no longer effectively perform all the types of work in this case and cannot be used interchangeably.

ESTABLISHING PRIORITY SYSTEMS FOR MAJOR JOBS

A priority system for major jobs is used to define the order of precedence in assigning the accomplishment of specific work orders involving construction, modification or rehabilitation. Normally, the method establishes subcategories such as:

(1) Top Priority
(2) Secondary Priority
(3) Routine Priority

Top Priority work requires immediate action, taking precedence over all other work to be scheduled. Normally, the assignment of a top priority is limited to work critical because of health, safety or security requirements or performance of operations essential to the organization serviced by the maintenance department.

Secondary Priority is assigned to work requiring expeditious handling to assure completion on a specific need date to support operations.

Routine Priority is assigned to work orders that may be accomplished on a first-in-first-out work schedule cycle where no specific need date has been established by the requester.

Regardless of the number of subcategories established, it should be recognized that any priority system for major jobs will involve a certain amount of rescheduling effort. A new job introduced into the system may take precedence over other jobs already scheduled or the need date on a job already scheduled may change to an earlier date. Also, jobs initially assigned a secondary priority may be reassigned a top priority when it becomes apparent that the requested need date would not otherwise be met. This constant necessity for shuffling is why scheduling is sometimes referred to as the challenge of matching work to be done with resources available to accomplish the work.

USING PHASE SCHEDULING ON MAJOR JOBS

In maintenance organizations responsible for performing a substantial amount of modification or rehabilitation work, formalized scheduling methods may be used to track each job. Each project is divided into increments or steps with a separate time frame for start and completion.

As an example, let us examine a project schedule divided into the following ten phases:

1. Feasibility Study
2. Budgetary and Planning Cost Estimate
3. Project Funding Authorization
4. Design
5. Detailed Cost Estimate
6. Design Review
7. Authorization of Construction
8. Construction
9. Acceptance of Work
10. Final Fit-Up for Occupancy

The following is a brief description of what occurs during each of these ten phases:

The **feasibility study** consists of a preliminary analysis of the scope of the job. The purpose of this phase is to determine generally what is to be done and what will be required to accomplish the modification or rehabilitation. This study is normally accomplished by a job planner or maintenance engineer in the maintenance planning office or by a planning team knowledgeable in the disciplines required to do the work.

The **budgetary and planning cost estimate** is a preliminary cost figure for accomplishing the project requirements. The estimate is a gross figure covering all work to be charged to the project plus a contingency amount for unforeseen occurrences. This cost figure is prepared by the individual or team that conducted the feasibility study.

The **project funding authorization** phase covers the time that will be spent in reviewing the feasibility study and budgetary and planning cost estimate for management funding approval to proceed with the work.

The **design** phase commences upon receipt of project funding authorization from appropriate management. This is sometimes called the architectural and engineering (A&E) phase. Construction plans and specifications are prepared for the job either in-house by the maintenance engineering section or contracted to an architectural and engineering firm. If the task is contracted, a maintenance planner should be assigned to provide liaison with the A&E firm. His responsibilities should be to track the progress of the job and provide coordination between the firm and the maintenance planning office.

A **detailed cost estimate** using the drawings and specifications is then prepared in the next step. Normally, if the plans and specifications have been contracted, the A&E firm also prepares the cost estimate. If such is the case, the detailed cost estimate should still be reviewed in-house by a construction cost estimator in the maintenance engineering section to assure that there have been no errors or omissions.

The **design review phase** consists of examining the construction drawings and specifications for errors, omissions or desired changes. The plans, specifications and the detailed cost estimate are then revised as required.

The **authorization of construction** phase consists of obtaining necessary signature authorizations to proceed with construction and release the plans and specifications. When the detailed cost estimate figure is greater than the budgetary and planning figure, signature authorization for the additional funding is often required before proceeding to the construction phase. Also, some maintenance departments issue a separate work order to authorize the construction phase. The purpose of this new work order for the next phase is to assure that no construction work is authorized or accomplished prior to completion of the design.

The **construction** phase is the next time period in the project schedule. This is sometimes called the brick-and-mortar step, since actual construction is accomplished in this phase. If the work is contracted in whole or in part, a contract construction inspector from the maintenance planning section is assigned to the project and contracts are let.

The **acceptance of work** phase follows construction. The job is checked for errors, omissions and faulty workmanship, any of which is documented on a "punch list." For contracted work, the list is prepared by a contract construction inspector. For in-house work, the list is prepared by a maintenance planner or construction coordinator. Discrepancies are then worked off and the work is accepted.

The **final fit-up for occupancy** phase consists of accomplishing items of work in connection with the physical occupancy of the space by users. The following types of work are often done in this last phase.

1. Drilling holes and pulling wires for installation of telephone instruments.
2. Wall installation of items as blackboards, bookcases, fire extinguishers, etc.
3. Room, corridor and building column numbering using paint stencils or tags.
4. Final site cleanup and floor finishing.

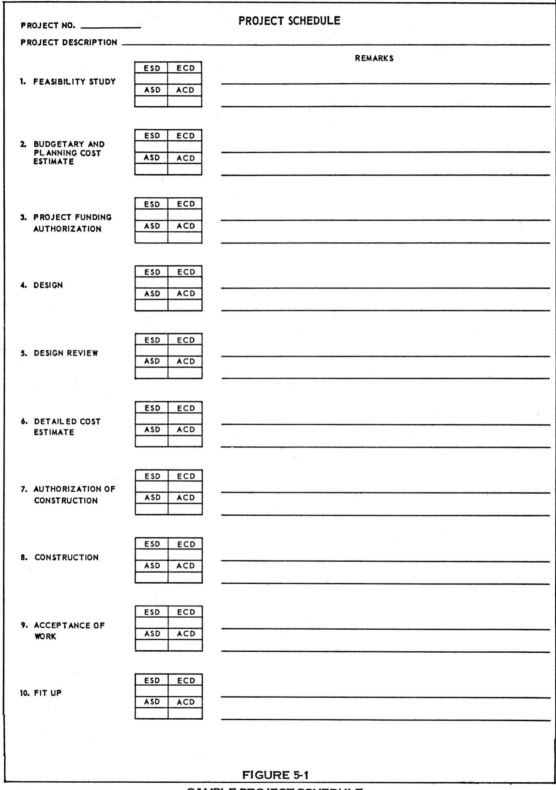

FIGURE 5-1

SAMPLE PROJECT SCHEDULE

Figure 5-1 is an example of a schedule format for this ten-phase approach. In order to use the schedule form as a control as well as a planning tool, note that for each phase there are blocks for posting actual start and completion dates as well as estimated dates. Blocks labeled "ESD" and "ECD" are used for posting estimated start dates and estimated completion dates. Blocks labeled "ASD" and "ACD" are used for posting actual start dates and actual completion dates for each phase.

APPLYING NUMERICAL PRIORITY SYSTEMS

In an effort to establish priorities for the accomplishment of work orders, some maintenance departments use a numerical formula. An example of such a formula is ABP. It works this way:

A = a numerical value for the priority assigned by the maintenance department
B = a numerical value for the priority assigned by a requester
P = the sum of multiplying value A by value B.

As an example, let us assume three would be the highest priority value and one would be the lowest priority value that could be assigned to a given job by both a requester and a maintenance planner. Thus, the highest priority (P) that could be assigned to a job would be a numeric value of nine (A=3, B=3, 3 x 3 = 9) and the lowest priority a value of one (A=1, B=1, 1 x 1 = 1). Assuming a maintenance planner assigned a value of three and the requester a value of two to a job, the resultant priority (P) would equal six (3 x 2 = 6).

The basic ABP method can provide a greater range of numeric priority values. For the sake of simplicity, the above example used values of only one, two or three. Use of a numeric range of one through five or one through ten provides for a larger number of numeric priority values.

A variation of the basic ABP method is to provide more numeric values for assignment by a maintenance planner as opposed to a requester or vice versa. This permits the individual with the greater range of values to have more influence on the numerical priority value assigned to the job.

Numerical priority systems provide a means for quantification. However, it should be recognized that the quantity value assigned by a requester or a maintenance planner is based upon individual judgment. The fact that this judgment is quantified does not guarantee its validity. Further, the method does not preclude the requirement for another judgment in ranking two or more work orders that are assigned the same numeric priority value.

HOW MUCH SCHEDULING?

The major thrust of work scheduling is to improve the efficiency of the maintenance department. The form of scheduling may range from the planning a foreman does in assigning craftsmen to the day's work all the way to an elaborate, centralized planning system employing specialists who allocate the time of each craftsman to a specific job. Within this range each maintenance department should select the point that provides optimum utilization of manpower. There should be as much planning as required for maximum overall efficiency as long as the scheduling system costs less than the cost of operating without it.

As a rule of thumb, a maintenance department employing more than a dozen men and more than three crafts should use more complex scheduling than just the day-to-day allocation of work by supervisors. This planning is necessary in order to coordinate the craft skills on larger projects and to assure an equitable distribution of work. As a maintenance department increases in size, the amount of effort spent on scheduling should also increase in order to maintain the efficiency level and quality of service.

There are practical limitations as to how comprehensive a scheduling system should be. A detailed schedule that becomes obsolete after the first few hours in a day due to emergencies is of little value. Some production plants have found a day schedule to be of value. For most operations, a weekly schedule proves to be the most effective because it is sufficiently flexible to handle most unexpected delays resulting from emergency work, weather, or other contingencies. If actual performance indicates around seventy percent adherence to schedules during normal operations, the value of the scheduling system is real. If the system does not demonstrate this level of adherence, it should be revised to get in touch with reality.

Another consideration in scheduling is the percentage of the total workload to be scheduled. The most effective systems recognize the inability of any maintenance department to anticipate all jobs, particularly emergency work. Thus, there is no attempt to schedule by work order the entire workload of each man. A simple allocation of some portion of the work force is made on a weekly basis for certain types of work such as trouble calls.

6 *Money-Saving Job Planning and Estimating Techniques*

Job planning consists of determining what is to be done and how it is to be done. Job estimating consists of determining the craft skills and materials required to accomplish what has to be done. These labor and material requirements may then be converted into dollars if a cost estimate is desired.

There are consultants in the marketplace who purport to have all the answers to the problems of job planning and estimating. Most of them seem to start with the assumption that the maintenance crew of the prospective client is comprised largely of dolts and laggards. They believe that the craftsmen and foremen are incapable of planning their work and that standards are required on all jobs in order to assure productivity.

These consultants then proceed to set their prospective client straight on how job planning and estimating should be done and how much money will be saved by applying their method. The major oversight in the approach of these consultants is that no two maintenance organizations are alike with respect to their mission, resources or facilities to be maintained. What works for one organization may be completely inappropriate for another. To a degree these consultants are in the marketplace with a solution (their method) looking for a problem. The problem they are looking for is an organization that is dissatisfied with its maintenance operation, is willing to buy consulting services and can be convinced that the consultant's method will work.

This chapter sets forth some of the most widely accepted techniques that have been used successfully in job planning and estimating. Their use, however, should be tailored to match the particular needs of the maintenance organization that endeavors to put them into practice.

PLANNING WHAT IS TO BE DONE

MODIFICATION WORK

Determining what is to be done on modification work is largely a matter of finding out what the requester thinks he wants and reducing this into needs. Wants are not always needs. The requester may want something that is not economically justifiable or even feasible. He may settle for something less, which will actually satisfy his needs.

One of the most counterproductive occurrences in modification work is to translate a requester's wants into design plans and specifications with a detailed cost estimate and then discover that he is unwilling to approve the construction because of the costs involved. The design is then reworked to reduce his wants to his needs in order to lower the costs of the project. The whole occurrence could be prevented if his wants had been reduced to needs before the design plans and specifications had been prepared. To accomplish this reduction, someone in the maintenance department should be assigned to the request for work. Sequentially, his job should consist of the following:

1. Confer with the requester.
2. Walk the area where the job is to be performed.
3. Reduce wants to needs by explaining the cost differences between the two.
4. Describe the needs in writing on a service request.

CORRECTIVE MAINTENANCE

When a piece of production equipment repeatedly breaks down, an investigation should be made to determine why this is happening. When the breakdown repeatedly occurs for the same reason, the cause can sometimes be eliminated by corrective maintenance. Rather than a major redesigning or rebuilding of the equipment, the corrective action may involve a relatively small change such as installation of a bearing of a higher load capacity, hard-facing a surface, fitting a larger grease cap or modifying a grease seal.

Job planning for this type of work starts by identifying the causes of the breakdown to determine if it always occurs for the same reason. Required corrective action can then be determined.

PERIODIC OVERHAUL

Planning of periodic overhaul can be accomplished using several methods. One method is to follow the manufacturer's recommendations for periodic overhaul based upon hours of operation. When these hours are converted to days, a plan can be devised to identify when the overhaul of each piece of equipment should be effected and what has to be done.

A second method that may be used to plan overhauls is to identify when a piece of equipment is repeatedly breaking down for a variety of reasons. The causes frequently indicate that the item requires rebuilding.

A third method of determining need for overhaul uses inspections and tolerance checks. This method is particularly applicable to machine tools. For example, if a machine can no longer hold prescribed tolerances, the odds are it needs new bearings.

TURNAROUND MAINTENANCE

In highly mechanized manufacturing plants and in many process industries turnaround maintenance is a vital function. For these types of operations it is essential that the equipment be shut down and overhauled in the shortest possible time. The most effective method of facilitating maintenance planning for this type of work is through close cooperation with manufacturing operations. If this cooperation is not established, the turnaround maintenance program is quite apt to be a dismal failure. When production schedules and maintenance schedules are not synchronized, two things will generally occur. First, manufacturing refuses to shut down in support of the maintenance requirements and their production schedules take priority over the schedules prepared by the maintenance department. Second, the maintenance department does not have the necessary manpower, materials and equipment available to do the job on an instant-response basis to the whims of manufacturing.

PLANNING HOW IT IS TO BE DONE

MODIFICATION WORK

Modification work consists of translating the requester's needs into a description of the work to be performed to meet those needs. The sequence of operations in performing the work must also be described. On major modifications the sequence consists of the following steps:

1. Site preparation including demolition
2. Structural work (rough carpentry)
3. Mechanical and electrical work
4. Finish carpentry
5. Painting

An example of lack of job planning occurred at one university. The job called for converting a complex consisting of three offices into three smaller offices with a reception area. The job order was released on a Monday. On the following Friday evening, the painters showed up and painted the existing office complex. The next Monday the demolition crew arrived at the site, disconnected the existing electrical fixtures and removed the interior walls of the complex. Thus, what should have been the final step in the project preceded the first step of site preparation. This happened because the craft foremen were not coordinating the sequence of their operations.

CORRECTIVE MAINTENANCE

For corrective maintenance to be accomplished, planning must occur as for modification work. First, special parts and materials must be placed on order and received. Lost time will be incurred if a craftsman is assigned a job and then spends half an hour or so discovering that the parts he needs have not yet been delivered. After materials have been received, required shop work should then be accomplished and, finally, field work.

PERIODIC OVERHAUL

Planning how overhaul work is to be done involves two basic decisions. First, a decision should be made as to whether some of the work should be contracted or if it can all be performed in-house; second, whether the work can be performed with the equipment in place or if it should be removed to a shop.

If the work is to be contracted and the equipment removed to the contractor's shop, in-house labor will still be required to disconnect the equipment. If this disconnect work is not properly planned, additional costs are usually incurred while the contractor's people wait around for the equipment to be disconnected so that they can take it out.

If the work is to be done in-house, it is necessary to determine which special parts need to be ordered and how the work will be divided between shop work and field work. For example, pipe should be cut and threaded in the shop and then installed in the field.

TURNAROUND MAINTENANCE

Because the length of the shutdown for performance of turnaround maintenance must be held to a minimum, it is essential that all of the work be thoroughly preplanned and organized.

Usually only the most highly skilled mechanics should be assigned to the jobs. This means that shutdowns and vacations must be coordinated. Without this coordination, the best men may not be available when the job is to be done.

Another costly oversight is failure to have all required special parts in stock. During disassembly or reassembly if a part or fitting is damaged the whole job may be delayed until a replacement is obtained. This delay can be avoided if all spare parts are in stock before the work commences.

Although these safeguards may appear to be obvious, they are not always employed and the results have been costly. One food processing plant in Los Angeles, California, had its mixing operation shut down for four days while a vital special fitting was being air-freighted from the East Coast equipment manufacturer. An original fitting that was badly worn disintegrated when a wrench was applied to it during disassembly. The damage was not the fault of the mechanics doing the job, but there was no spare stock and practically an entire week of production was lost for want of a $26.00 fitting.

ESTIMATING

DETERMINING CRAFT SKILLS AND MATERIALS

The description of the work to be done provides the basis for determining material and craft skill requirements. On modification jobs the design plans and specifications provide the basis for determining material and craft skills requirements. On smaller jobs this function is performed by the person who does the job walk and planning.

PREPARING THE ESTIMATE

A proficient cost estimator is a highly knowledgeable technician. He must know:

- How to determine the amount of material required
- How to price materials to convert material requirements to dollars
- Which craft skills are required to perform the various elements of work
- How to use appropriate standards to determine how many labor hours are required for each craft skill
- The labor rates to apply to the various craft skills to convert hours to dollars

Properly utilized, a detail cost estimator can actually plan the sequence of the job, since his estimates for craft labor are based upon a certain sequence of work performance. If the method he devised in preparing the cost estimate is not essentially followed, there is quite apt to be a marked variance between the cost estimate and the actual costs.

The actual costs of a job are sometimes open to criticism for being too high. These high costs are usually attributable to poor job planning. The cost estimate is an integral part of the job plan. It should not be documented on the back of an envelope. A prescribed form should be used to record a cost estimate, which should then be filed for future reference in controlling job costs.

Figure 6-1 is an example of a detail cost estimate sheet that contains the essential elements of information of such a form. Note that in the upper right-hand corner there are specific spaces provided for recording the job number, the number of sheets contained in the cost estimate, the preparer, and the date. The upper left-hand portion of the form provides space for identifying and describing the project. The body of the form is used for identifying each item of work. There are spaces for quantity, unit price and costs of materials. Labor data identifies the craft, total man-hours, craft rate and cost. There is also a space for identifying the cost estimate of contracted work for the job. The order in which items are listed on the form essentially describes the sequence of operations that should be followed in performing the work.

CONSTRUCTION PLANNING AND COORDINATION TECHNIQUES

Various methods may be used for reviewing and coordinating construction projects. The following are some of the more successful ones.

PERFORMING JOB WALKS AND PLANS REVIEW

On large modification projects involving several crafts, the front-line supervisor responsible for performing the project should perform a job walk and review the plans,

COST ESTIMATE DETAILS

PROJECT _____

DESCRIPTION OF WORK _____

FSR NO. _____

SHEET ____ OF ____

BY ____ DATE ____

ITEM	DESCRIPTION	QUANTITY	MATERIALS			LABOR					CONTRACT	TOTAL
			UNIT PRICE	AMOUNT		CRAFT	TOTAL M/H	RATE	AMOUNT			

FIGURE 6-1

SAMPLE COST-ESTIMATING SHEET

specifications and detailed cost estimate prior to the commencement of work. This action provides for any clarification of requirements in the plans and specifications. It also permits any required adjustment to the cost estimate to be made before the job starts.

USING THE MONDAY MORNING APPROACH

This technique is usually used when each craft foreman is responsible for planning his part of a modification job involving several crafts. The method provides for a specific time when all jobs in-work or going into work are reviewed jointly by the foremen to assure proper interface in the sequence of the operations. The technique gets its name from the fact that the meeting occurs at the beginning of the work week. Mondays are selected because they permit a status review of jobs currently in-work and, particularly, a review of what was accomplished during the weekend. The work to be done in the coming week is then reviewed and coordinated among the various craft foremen.

APPLYING THE CONSTRUCTION COORDINATOR METHOD

This method is also generally used where each craft foreman does the job planning for his craft. The role of the coordinator is to release the work and assure that it is performed in the proper sequence. He also serves as a liaison to the design group regarding plans and specifications for the job, coordinates any contracted work at the job site and processes completion paperwork. In some instances, the Monday morning method is used in conjunction with the construction coordinator approach in an effort to gain the advantages of both methods.

USING A PLANNING STAFF

When using a planning staff, the responsibility for job planning and estimating is assigned to a separate staff unit. Personnel within the unit are usually called maintenance planners, but their individual responsibilities may vary. Some planning units have maintenance planners assigned and specialized by a given craft skill. When this is the case, the coordination of the sequence of operations for a job is performed by another individual on the planning staff usually called a job coordinator or construction coordinator. In other instances, the maintenance planner is a generalist who coordinates the sequence of operations and does the job planning for all the craft skills required for a given job.

The purpose behind the creation of a planning staff is to attempt to assure proper planning of work through specialization. Inherent in the line-and-staff concept of organization is the potential conflict between the two parts. In the pressure cooker atmosphere of construction modification work, this conflict is quite apt to erupt frequently. The line personnel consider the staff operation to be superfluous and the staff personnel regard the line operation as being comprised of a bunch of uncooperative ninnies. Despite the recognized disadvantage of having the job planning performed by a staff organization, large-scale maintenance organizations employing over 200 craftsmen or with an average workload of over 100 active construction job orders have used it successfully. When a department gets that big, specialization is usually warranted and proves to be the most effective way of getting the planning accomplished.

A key to the successful use of a planning staff is to insure that orders are transmitted upward only. Staff units will not operate effectively when they hand orders down to men in line operations.

7 *Implementing Periodic Maintenance Systems*

This chapter uses the term "periodic maintenance" rather than "preventive maintenance" because a comprehensive periodic program encompasses maintenance operations that go beyond the prevention of equipment breakdown. Properly planned, a periodic maintenance program can effectively identify who, what, when, where, why and how specific repetitive operations are to be performed. Manpower, material and maintenance equipment needs can be specifically identified and justified. These requirements can then be expressed in terms of material and labor requirements and dollars for budgetary planning purposes.

A comprehensive periodic maintenance system is vital to the proper planning and justification of this type of required maintenance effort. It permits a maintenance manager to describe how his department plans to accomplish its mission. This chapter presents facets of proven scheduled maintenance systems that can be used in establishing or improving your system.

WIDENING THE SCOPE OF A PERIODIC MAINTENANCE SYSTEM

Often the scope of a periodic maintenance system is restricted to preventive maintenance of personal property items such as machine tools in a manufacturing operation. It is possible, however, to expand a system to schedule repetitive inspection and maintenance of building equipment, utility distribution systems, janitorial services, groundskeeping and gardening activities, and the painting of buildings.

To accomplish all of these types of periodic maintenance, specific work orders should be prepared. Where several kinds of maintenance are required for a given item, a separate schedule is established for each operation. For example, a machine tool may have separate schedules for lubrication, electrical checks, mechanical checks and tolerance checks. For each of these activities, a separate work order is generated at the prescribed time and distributed to the appropriate craft skill.

All maintenance and repair activity for an equipment item should be reported for inclusion in an historical record. A single record that contains the complete maintenance history of the item, including preventive maintenance, breakdown, repair and rebuild data should be maintained.

ESSENTIAL INFORMATION FOR A
PERIODIC MAINTENANCE SYSTEM RECORD

The minimal information required to establish a record in a system should consist of the following four items:

1. An identification number for the piece of equipment or work activity (what)
2. The location of the maintenance work (where)
3. A determination of what operations are to be performed and how often (why, when and how)
4. A judgment as to what craft skill is required to perform the operation (who)

THE IDENTIFICATION NUMBER

The purpose of the identification number is to give individual identity to an item or work activity upon which periodic maintenance operations are to be performed. The use of the identification number varies according to the type of maintenance being performed. For example:

Plant Equipment—A number is assigned to each individual piece of capital equipment such as machine tools or office equipment.

Building Equipment—A number is assigned to each individual item of equipment such as a boiler, air handling unit or a booster pump in the water distribution system.

Relamping—A number is assigned to each given zone or area in a facility where relamping is to be performed on a scheduled basis.

Painting—A number is assigned to each given zone or area in a building where painting is to be performed on a periodic basis.

Janitorial Operations—A number is assigned to each given zone or area in a building where floor refinishing, carpet cleaning, etc. is to be accomplished on a periodic basis other than that of daily housekeeping.

Grounds and Landscaping—A number is assigned to each type of groundskeeping or landscaping operation that is to be performed on a repetitive basis and for which a separate accumulation of labor hours and material costs is desired.

DESIGNATING LOCATIONS FOR THE WORK TO BE PERFORMED

A location designation tells the craftsmen where the work is to be performed. For plant or building equipment the location may be identified by using a building and room number. For relamping or painting the location can be cited by a building number and zone identification.

DETERMINING OPERATIONS TO BE PERFORMED AND THEIR SCHEDULES

The operations to be performed and the schedule for their performance will vary according to the type of maintenance. For example, for a given machine tool assigned an identification number, the operations to be performed may include oiling, an electrical check and a tolerance check. Each of these operations may be performed on a different schedule. Based upon the manufacturer's recommendations and experience, oiling should be performed monthly, an electrical check quarterly and a tolerance check every six months. The schedule should then be established for each operation based upon the number of weeks in the cycle (i.e., oiling four weeks, electrical check twelve weeks and tolerance check twenty-six weeks). The week for the first work order is then selected. Thereafter, the work orders are prepared based upon the selected time cycle.

Other types of maintenance may require different arrangements. For grounds-keeping a work order may be issued annually for tree pruning in a given area, which should be identified by an identification number. In this instance there would be only one scheduled operation for the identification number.

For painting operations the schedule may vary, based upon the type of area. Corridors in an office building may be scheduled for painting every twelve months, while the offices may be painted only every four years.

IDENTIFYING THE CRAFT SKILL REQUIRED

For each operation it is also necessary to determine the craft skill required. For example, an electrical check requires a trained electrician and a tolerance check requires a machine tool mechanic. For each scheduled operation the required craft skill should be identified in order to assure proper assignment of the work.

ELEMENTS OF A PERIODIC MAINTENANCE SYSTEM

In establishing a periodic maintenance system, the following six items are primary considerations:

1. Planning requirements
2. Preparation of work orders
3. Distribution of work orders
4. Following-up
5. Collection and posting of cost and service data
6. Integration of data

PLANNING REQUIREMENTS FOR PERIODIC MAINTENANCE

Planning requirements involves determining what maintenance work should be performed on a periodic basis and establishing a schedule for each item of work. This planning would include establishing the need for electrical, mechanical, and lubrication checks on machine tools, and scheduling relamping or painting activities by area within buildings. Manufacturer's brochures can be used to determine the need for and frequency of maintenance activities for equipment items. For other maintenance activities, the judgment of foremen experienced in that type of maintenance can be used.

Planning of periodic maintenance should be a continuous process. Maintenance history should be reviewed periodically to determine if the operations and their schedules should be changed. This follow-up evaluation can result in cost savings. For example, when equipment items are under warranty, initial scheduled maintenance is set according to the manufacturer's instructions to preserve the warranty provisions. However, actual use of the equipment may reveal that the time between maintenance operations may be lengthened without impairing the operation of the equipment since the manufacturer prescribed a schedule which would ensure that his equipment performed at the optimum level under all uses. The actual use of equipment in a given plant may permit change in the maintenance cycle without impairment of the equipment. One plant was able to successfully increase the time between greasing and packing of the water pumps from four to twelve weeks. This action yielded a savings of $3500 annually.

PREPARING WORK ORDERS

After it has been determined what operations are required, how often they are to be performed and what craft skill is required to do the work, it is often necessary to plan for the preparation of paper work. Work orders can be prepared on an automated or manual basis. Regardless of which method is used, work orders should be prepared in advance of the scheduled dates for performance of the operations. If the paper work authorizing the work is not available for distribution, the schedules become meaningless.

DISTRIBUTING THE WORK ORDER

Work orders should not merely be given to the craftsman. They should be distributed by the assigned foreman to leadmen and craftsmen with due consideration given to the amount of work being released. The orders should be grouped by geographical area or building to reduce travel time. In essence, this grouping of work orders provides a practical route for accomplishing the work. The elimination of duplicate travel time can result in substantial savings. For one school district it eliminated over $40,000 annually in travel time.

FOLLOWING-UP

There should be a method for identifying work orders for which no completion notice has been received so that corrective action can be initiated. Identification of these work orders may be accomplished manually using an "open order" file where a reference copy of each work order issued is filed until a completion notice is received.

Another aspect of following-up work orders is a comparison of the actual performance time with an assigned standard. When there is a marked difference between the two times, the foreman should investigate to determine the cause. Usually a marked difference occurs because either the operations were not performed properly or because some other work was done that was not noted on the work order. In the former instance, the craftsman may require additional instruction. In the latter instance, the additional work done should be investigated to clarify what really occurred.

COLLECTING COMPLETED WORK ORDERS AND POSTING COST AND SERVICE DATA

An integrated maintenance system requires more than completion of a given item of work. The completed work order should be recorded for the purpose of closing it out in the "open order" system, maintaining an historical maintenance record on an item, and identifying costs for material and labor expended.

INTEGRATING THE PERIODIC MAINTENANCE DATA WITH OTHER WORK

To determine the effectiveness of the periodic maintenance activity for an item of equipment, the total maintenance required on that item must be considered, including breakdown, repair, rebuild effort, and scheduled maintenance. A complete history of breakdown service may indicate that the types of preventive maintenance being performed or the schedule for their performance require revisions. Analysis of a complete maintenance history also allows judgments to be made regarding retention and future use of equipment.

Once the planning process has identified all the periodic operations that should be performed, cost savings can be realized by the standardizing maintenance materials. The variety of materials in stock can be reduced and the items bought in quantity. One company achieved an annual savings of $20,000 by standardizing lubricants used on their manufacturing equipment.

DOCUMENTS FOR A SCHEDULED MAINTENANCE SYSTEM

Certain documents are an essential part of a periodic maintenance system. These documents are:

1. The Inventory Sheet
2. The Operations Sheet
3. The Work Order
4. The Historical Record

USING AN INVENTORY SHEET

The purpose of the inventory sheet is to provide data on the item for which periodic maintenance is to be performed. Figure 7-1 is an example of an inventory sheet for equipment items. Note that the top portion of the form provides spaces for recording such data as purchase price, acquisition date, identification (key) number, location and nomenclature. Space is also provided for recording manufacturer's

EQUIPMENT INVENTORY SHEET

OWNER CODE	PROPERTY NUMBER	REQ. CCC	EQUIP. TYPE	PURCHASE PRICE	ACQ. DATE MO. YR.	KEY NUMBER

LOCATION BLDG. ROOM	OTHER	NOMENCLATURE

MANUFACTURER'S DATA

NAME	SERIAL NUMBER	MODEL NUMBER

DRIVE BELTS		BEARINGS	
QUANT.	PART NUMBER AND NAME	QUANT.	PART NUMBER AND NAME

ELECTRIC MOTOR DATA

NAME OF MANUFACTURER _____   SERIAL NUMBER _____   MODEL NUMBER _____

MOTOR TYPE _____   CODE _____

H.P. _____   PHASE _____

VOLTS _____   AMPS _____

CYCLES _____   RPM _____

FRAME _____   DESIGN _____

RATING _____   SERVICE FACTOR _____

FRONT BEARINGS _____

OPPOSITE SHAFT END _____

SPECIFICATIONS _____

SHAFT SIZE _____   POWER SOURCE _____
BUILDING   ROOM   PANEL   CIRCUIT

AUXILIARY ITEMS

STARTER: MAKE _____

MODEL _____ SIZE _____

ENCLOSURE TYPE _____

THERMAL HEATER:

SIZE _____ TYPE _____

DISCONNECT: FUSE ☐   BREAKER ☐

SIZE _____ MAKE _____

MODEL OR TYPE _____

DATA ON ACCESSORY AND AUXILIARY ITEMS

_____
_____
_____
_____
_____
_____
_____
_____
_____
_____
_____
_____

PREPARED BY: _____
NAME   DATE

FIGURE 7-1
SAMPLE INVENTORY SHEET

identification data to facilitate correspondence with the manufacturer regarding the item. Space is provided for data on spare parts (drive belts and bearings), electric motors, and accessory and auxilliary items. An inventory sheet should be filled in immediately upon receipt of an item, while the information on the tags affixed to the equipment by the manufacturer is still completely legible. Other information may be obtained from a manufacturer's brochure on the item.

The inventory sheet comprises a ready reference for ordering spare and replacement parts.

APPLYING AN OPERATIONS SHEET

The main purpose of the operations sheet is to provide the craftsman with a reference document regarding the specific maintenance to be performed. Figure 7-2 is an example of an operation sheet. Note that the first three lines on the form list the item upon which the work is to be performed, the craft skill required to do the work, the location of the item, the number assigned to the operation, and the identification (key) number for cross-reference to the inventory sheet. The schedule for the operations is then indicated, followed by a space to describe the specific operations to be performed. A space is provided for describing equipment and lubricants to be used in performing the work. The bottom portion of the form provides space for indicating subunits, availability of drawings and manuals, and assignment of a labor standard.

DESIGNING A WORK ORDER

A work order is used to satisfy the following four purposes:

1. To serve as a notice that the periodic operation is to be performed
2. To provide authorization for the expenditure of labor and materials in performance of the work
3. To furnish a document for recording that the operation was performed
4. To provide a document for the written feedback of other information such as materials used, other work which was or may be required, etc.

The information on the work order should be standardized so that craftsmen know where to go, what to do, and how to report the work accomplished. In designing the work order, needs of the individuals using the forms should be considered. Suitable space should be provided for manual entries. The number of entries should be minimized since most craftsmen do not enjoy filling out lengthy forms.

The need for the collection and posting of cost and service data should also be considered when the work order form is designed. The size and number of copies required should provide for a follow-up system.

THE HISTORICAL RECORD

The historical record is the document that provides for the integration of data. The record may be automated or manual. In either instance it should contain the

OPERATION SHEET FOR SCHEDULED MAINTENANCE

DATE: _____

SUPERSEDES: _____

| NOMENCLATURE OF UNIT: | CRAFT CODE: |
| LOCATION: | OPERATION NUMBER: |

| KEY NUMBER | JOB NUMBER | ID CODE | |

SCHEDULED FREQUENCY: Indicate in weeks (e.g. 4 times a year would be 13 weeks) ⟶ ☐

OPERATIONS TO BE PERFORMED

_____
_____
_____
_____
_____
_____
_____
_____
_____
_____
_____
_____
_____
_____
_____
_____
_____

EQUIPMENT/LUBRICANTS USED

_____
_____
_____

SUB-UNITS

Are there sub-units in connection with the above unit? NO ☐ YES ☐ If yes, state nomenclature, location of sub-units and if it is mechanical or electrical equipment.

_____
_____

ARE AS-BUILTS OR MANUALS AVAILABLE? NO ☐ YES ☐ If yes, who has them? _____
NAME   PHONE EXT.

LABOR STANDARD DATA

Labor Standard: Indicate the Labor Standard to the nearest tenth of an hour. ⟶ ☐

Is the Labor Standard to be printed on Maintenance Work Orders in the "FUNCTION" block? NO ☐ YES ☐

FIGURE 7-2

SAMPLE OPERATION SHEET

86

complete history of all work performed including, when applicable, scheduled work, breakdown of equipment, repair and rebuilding of equipment.

The type of information retained is important. For example, if belts repeatedly have to be changed on a certain motor drive or the motor was excessively rewound, corrective maintenance may be required. On the other hand, to record that a forklift truck was driven into the gear unit and broke off the foot may not be of any particular value.

APPLYING PERIODIC MAINTENANCE

Now that we have discussed the four essential information items, the six elements, and the four documents that should be part of a periodic maintenance system, let's look at applying a program. The practical use of a periodic maintenance system will vary according to the mission of the particular maintenance department. When the work is essentially preventive maintenance, the program should be set up on the basis of reducing overall operating costs. Preventive maintenance is a means to an end, not an end in itself.

One electric power distribution agency instituted a power pole inspection program. The maintenance men climbed over 20,000 poles in one year. They found work to do on less than twenty poles. On the other poles, the men just got exercise. The whole program didn't save a dime and a periodic visual inspection from the air or ground would have been just as effective.

In production plants, the shutdown of some equipment for no other reason than periodic inspection and adjustment may be inadvisable if production is to be maintained. It should be recognized that it is sometimes more profitable to the organization as a whole to run a piece of equipment until it breaks down rather than deliberately interrupt a pressing production schedule. While this may be hard for maintenance people to accept, it is something that should be considered if no threat to the safety of personnel or adjacent equipment could result from the unexpected failure.

Figure 7-3 lists some potential building equipment candidates for periodic maintenance. Note that the word "potential" is used. Periodic maintenance on equipment items should be tailored to fit the function of the specific item rather than applying the same operations to all equipment. For example, the cost of periodic inspections of small electric motors on ventilating fans can readily exceed the cost of unit replacement at the time of a failure. On the other hand, the large, primary air handling units should be inspected and maintained.

Figure 7-4 indicates the major considerations and equipment classifications used by one large operating division of a major corporation in selecting equipment for periodic maintenance.

In integrated manufacturing lines, such as oil refineries, preventive maintenance is justifiable because the investment in production capacity is much greater than the cost of the individual instruments themselves. Also, many of the key pieces of equipment, such as instruments, are usually readily accessible for inspection.

CONSIDERATIONS FOR PERIODIC MAINTENANCE OF EQUIPMENT

Heating, Ventilating and Air Conditioning (HVAC) System

Air Handling Fans and Motors
Condensing Unit—Pumps and Fan Motors
Hot Water Boilers
Boiler Pumps and Motors
Air Filters
Water Circulation Pumps and Motors

Domestic Water System

Hot Water Boilers
Hot Water Circulation Pumps and Motors
Chilled Drinking Water System Pumps and Motors
Urinal Flushing Systems

Fire Protection and Safety Systems

Emergency Lighting System
Fire Detection Devices and Alarms

Internal Transport Systems

Elevators
Escalators
Dumbwaiters
Compressed Air Door System Pumps and Motors

Electrical Distribution Systems

Lighting and Ballasts
Panels and Breakers
Transformers

FIGURE 7-3

**EXAMPLES OF POTENTIAL BUILDING EQUIPMENT
FOR PERIODIC MAINTENANCE**

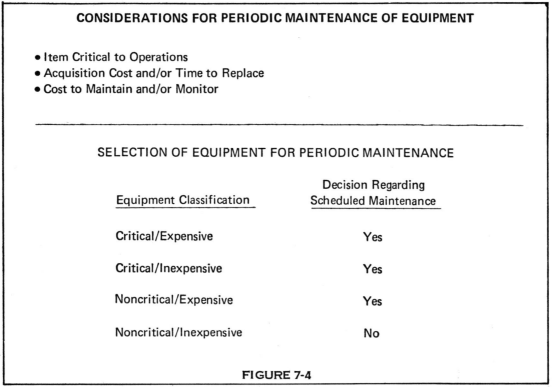

CONSIDERATIONS FOR PERIODIC MAINTENANCE OF EQUIPMENT

- Item Critical to Operations
- Acquisition Cost and/or Time to Replace
- Cost to Maintain and/or Monitor

SELECTION OF EQUIPMENT FOR PERIODIC MAINTENANCE

Equipment Classification	Decision Regarding Scheduled Maintenance
Critical/Expensive	Yes
Critical/Inexpensive	Yes
Noncritical/Expensive	Yes
Noncritical/Inexpensive	No

FIGURE 7-4

CONSIDERATIONS AND SELECTION OF
EQUIPMENT FOR PERIODIC MAINTENANCE

One chemical plant tempers ideal preventive maintenance with the practical considerations of a continuous operation by taking advantage of a breakdown. While the component that broke down is being repaired, other vital inspections and replacements are also accomplished. Use of this technique is most successful where deficiencies are observed and recorded during operating inspections. When the breakdown occurs, craftsmen can move in quickly and work until the entire job is done. Usually production people can be convinced of the need for taking a few more hours for the additional work at the time of a breakdown.

For janitorial operations periodic maintenance schedules may be assigned to any repetitive operations beyond daily housekeeping activities, such as floor refinishing, carpet and drape cleaning, and periodic cleaning of planters, air diffusers, light fixtures, and exterior windows.

Combining various operations can also be cost effective. An example is combining the cleaning of light diffusers on fluorescent lighting fixtures with the relamping program. The labor time expended in the removal and replacement of the diffuser is applied to both the cleaning activity and the tube replacement. A combined operation is particularly useful in areas used by the public where image is given a high priority, such as in airline terminals. In one airline terminal building combining of activities resulted in savings in labor in excess of $28,000 per year.

8 Designing Realistic Maintenance Requirements into Facilities

Most maintenance managers will agree that maintenance requirements are almost invariably not given adequate consideration when facilities are being designed. The reasons given for this neglect vary, but the following represent a general consensus of opinion by those most prone to criticize.

1. Architects are half-artist and half-engineer. The artistic half has a strong desire to express itself. The appearance of a facility frequently takes precedence over its utility.
2. Electrical and mechanical engineers are often more concerned with the fascinating designs they contrive than they are with the future operation and maintenance of their building systems.
3. The maintenance manager in an organization is usually so far down in the management hierarchy that no one even considers his participation in design review.
4. Even when some consideration is given to maintenance during the design concept phase, changes in design during construction often nullify the original plans. The construction contractor is interested in getting the construction accomplished in a manner that will assure him the most profit, and he is not concerned with maintenance and repair requirements after the expiration of his warranty.

In short, no one involved in the design really seems to pay much attention to the maintenance and operation requirements for the facility since these are not going to be their challenges.

In order to avoid unnecessary maintenance and operating costs, the maintenance department should play an active part in the planning phase of a new facility. It is up to the maintenance manager to convince his management to permit his participation. If he fails to make the request through indolance or apathy, he is not performing his total job.

Some of the items which should be considered during the design phase are:

- Windows
- Janitorial Closets
- Power and Water Sources
- Storage Areas for Maintenance Tools and Supplies
- Lockers for Maintenance and Janitorial Personnel
- Building Hardware and Fixtures
- Logistics Operations
- Location of Building Equipment

HOW MANY WINDOWS AND WHERE?

Many architects like to have a lot of glass in the buildings they design. It looks nice. It also can cost a lot of money for heating, cooling, and keeping the glass clean.

CLEANING GLASS

Two basic methods are usually available for window cleaning—"over the roof" and "up from the ground." In both cases, special equipment is going to be required for any structure above one story.

How much special equipment is needed depends on how the structure is designed. The cost can be as low as $15 for a common garden hose and a squeegee with a six-foot handle to as high as $50,000 for a scaffolding rig for "over the roof" work. Set-up time can vary from five minutes to fifteen hours. Plans for installing water and electrical power sources for the window washing equipment should be instituted during the construction of the facility.

How much cleaning will be required is dependent upon many variables such as weather (rainfall), the surrounding environment (amount of smoke and dust), and how dirty you are willing to let the glass get (quality of maintenance effort).

When the building is being planned, window washing requirements should be considered. Accessibility is a primary consideration. When the architect plans to have glass in the structure, consideration should be given as to how a window washer is going to get to it so the costs for special window washing equipment can be readily identified, water and electrical power sources planned, and repetitive labor costs for site setup lessened.

HEATING AND COOLING

Glass is a poor insulator. As the costs for energy continue to increase, the propensity to construct structures with a large amount of glass in the walls should decrease. This is because the money spent for heating and cooling a building is becoming a cost of substantial magnitude. Realistic maintenance/mechanical engineering during the architectural concept stage can substantially reduce the heating, ventilating, and air conditioning costs of a future structure.

Energy consciousness is becoming important. Conservation means planning for the future in new ways. The message to stop waste has been heard. Energy can be conserved by more efficient heating and cooling systems and building insulation.

You can apply a complete maintenance program to the heating, ventilating, and air conditioning system of a facility. Chilled water and reheat and preheat coils can be constantly cleaned and protected against freeze-up in the winter, the static pressures of filters can be constantly checked to determine cleaning and replacing schedules, pumps can be adjusted to control leakage and shaft erosion and motor maintenance can be performed to eliminate dirt and moisture, but all this maintenance work cannot change the fact that a system was poorly designed in the first place. Half the **energy** used in most existing heating and cooling systems is usually wasted.

PLANNING JANITORIAL CLOSETS

HOW MANY?

In determining the number of janitorial closets required in a facility several variables should be considered. These are:

- The number of floors in the facility
- The type of floor finishes to be maintained
- The proposed use of the areas
- The number of restrooms in the facility

Some general rules in planning are:

- There should be at least one janitorial closet per floor.
- There should be at least one closet per restroom complex.

Location next to or in restrooms is ideal for janitorial closets for two reasons. First, the restroom provides the water source and sewage system for the mop sink. Second, storage of restroom supplies is within the immediate area, which facilitates dispenser replenishment.

WHAT SIZE?

The size of the janitorial closet should be determined by several considerations such as:

- The type of equipment that needs to be stored
- The amount and types of supplies that need to be stored

The amount of storage space provided for equipment should be determined by what type of floor finishing will be in the area. Tile floor will require scrubbing equipment, and adequate floor space should be provided in the closet for equipment storage. Carpeted areas will require storage floor space for vacuum cleaners.

The amount and types of cleaning supplies that will require storage depends largely on the use of the area being serviced. For example, office areas require dust cloths, upholstery and fabric cleaners, furniture polish, and glass cleaner, while manufacturing areas require shop cloths, solvents, and mops. The amount of shelving required for storage is not only dependent on what type of supplies need to be stored, but also how many. The quantities to be stored should be based upon anticipated usage. Areas being serviced on a seven-day cleaning schedule will use 35 to 40 percent more supplies than areas on a five-day schedule.

An example of a failure to provide janitorial closets of a proper size occurred in an office building in Washington, D.C. The janitorial closets were designed with only sixteen square feet of floor space, most of which was occupied by a mop sink. As a result, most of the supplies and all power cleaning equipment had to be stored in the basement. Carting the supplies and equipment each night on a five-day housekeeping schedule amounted to thirty hours per week or three-fourths of a man-year in labor, which cost over $9000 annually in labor and related payroll expense. But the story does not end here. Due to pilferage of supplies and the theft of some equipment, a lock-up room had to be constructed in the basement. This extra room cost an additional $2500.

PLANNING POWER AND WATER SOURCES

ELECTRICAL

Power sources for cleaning equipment are often overlooked in the electrical design of a facility, particularly in public areas such as hallways and lobbies. As a result, additional wiring and electrical outlets must be installed or miles of extension cords provided so interior janitorial services and exterior window washing can be accomplished. Lighting and electrical outlets in electrical and telephone rooms should also be provided to facilitate maintenance and repair work.

WATER OUTLETS

Besides the water supply in janitorial closets, additional sources should also be considered, particularly for exterior cleaning of sidewalks and entrances as well as window washing. Water outlets should also be provided in equipment rooms for cleaning operations.

STORING MAINTENANCE TOOLS AND SUPPLIES

In addition to storage space for janitorial and restroom supplies, some lock-up areas should be planned for maintenance tools and supplies. Building maintenance requires light tubes and ballasts, filters for air conditioning equipment, equipment lubricants, plumbing supplies, etc. Not a lot of space is needed, but there should be some maintenance supply rooms provided in a facility and they should be located near the equipment being serviced. Otherwise, there will be a loss in maintenance labor hours due to nonproductive travel time in getting the necessary tools and supplies to do the job.

LOCKERS FOR PERSONNEL

Locker space should be provided for janitorial and maintenance personnel. Without lockers there will be pilferage of hats, coats, jackets, purses, lunch buckets and other personal property. Lockers do not take up much space and are well worth the investment to head off labor grievances.

STANDARDIZING BUILDING HARDWARE AND FIXTURES

One of the biggest problems in a multibuilding complex is to have a variety of locks, door hardware, bathroom fixtures, floor finishes, and window blinds. The level of effort required to procure, replace, and repair parts in inventory caused by this variety can be staggering. Standardization of these items in a site development saved one company an estimated $75,000 in maintenance stock inventory costs, not to mention the reduction in buying effort resulting from having fewer line items in stock.

FACILITATING LOGISTICS OPERATIONS

The term "logistics operations" pertains to the handling of furnishings, equipment, and materials within the facility. The design should facilitate these operations.

Where heavy volume is anticipated, freight entrances and loading docks should be provided with adequate space to maneuver delivery vehicles. Freight elevators should have a weight capacity and cab size suitable for the intended use of the facility. In one office building in Washington, D.C. freight elevators were provided, but the cab height was only seven feet. To move large desks and sofas the roof had to be removed from the cab. The removal of the roof solved the problem, except for at the top floor, where the clearance between the top of the cab and the roof was only 18 inches. Because of this height limitation, the furniture had to be unloaded on the floor below and carried up the stairs to the top floor. The mention of stairs leads us to the next topic.

Stairways and landings should be adequate for moving furniture and equipment from floor to floor. This is of particular importance in office buildings in which there are no freight elevators and which have passenger elevator cabs too small to accommodate large items of furniture. One comptroller moving into a new building had to buy smaller desks for his entire office force because it was not feasible to move them up to the floor in the new building from the inside. If windows and casings were removed and a high-rise crane rented, the desks could have been loaded in from outside the building, but the costs involved ruled out this solution.

Trash rooms should be large enough to handle standard trash containers. One multistory office building in Los Angeles, California was designed with a trash room on each floor. The rooms, however, were too small to accommodate standard trash bins. To rectify the situation, special bins had to be fabricated, at a cost in excess of $16,000, to handle trash removal.

LOCATING BUILDING EQUIPMENT

Maintenance engineering can result in significant cost savings in the mechanical design of facilities. These savings largely accrue from adequate consideration of the needs for accessibility to equipment for service, repair, and replacement. As examples of poor design for maintainability consider the following:

1. Induction boxes in an airport terminal building's heating, ventilating, and air conditioning system were jammed into ceiling areas with no consideration being given to service requirements. As a result, eight to ten labor hours were required each time access to a box was necessary for adjustment or repair. On the average this increased maintenance and repair labor costs by over $30,000 annually.
2. In the same facility, elevator equipment rooms were situated so that the only access was through the back panel of the cabs. To remove and replace a back panel took one hour of labor. This problem increased elevator service maintenance costs by $3000 annually.

Over twenty years ago field service maintenance personnel began to participate as part of the design team for the development of an aircraft. Their job was to assure that maintainability was built into the airframe. In comparison, we have a long way to go in designing facilities. Brick and mortar has a life expectancy substantially longer than aircraft so the costs for increased maintenance and repair labor resulting from poor design are substantially higher.

FINISH ITEMS

As the architectural design progresses, it is necessary to make judgments on finish items for floors, walls, and ceilings. The type of finish items selected should be based upon the criteria of:

- Planned use of area
- Installment cost per square foot
- Service life (in years)
- Maintenance cost

ANALYZING PLANNED USE

The planned use of the area should encompass the following considerations. First, is the area to be used for manufacturing, laboratory, clean room, office, public service, or what? Second, what type of traffic is expected, i.e., pedestrian, mobile equipment, or both? Third, will the traffic be heavy, medium, or light.

INSTALLED COST PER SQUARE FOOT

The estimate of installed cost per square foot includes both labor and material costs involved in the installation. In a protracted study, a replacement installation cost would also be included. For example, in comparing a vinyl asbestos tile floor with an average life of sixteen years to nylon carpeting with an average age of eight years, the second installation cost of the carpeting would be included in the sixteen-year study period or planning horizon.

COMPUTING SERVICE LIFE

The service life estimate should be tempered by the planned use of the area, since heavier traffic normally results in a shorter life. On floor surfaces for example, carpet in heavy foot traffic areas such as public lobbies will normally last less than half as long as the same carpet in a light traffic area such as a corridor in an executive area.

ESTIMATING MAINTENANCE COSTS

An analysis of maintenance cost takes on several dimensions. First, there is the factor of maintenance labor costs, which depends upon whether the area maintained is obstructed or unobstructed. For example, wall cleaning or painting in corridors is much faster than in obstructed areas such as offices. Labor cost is also affected by the traffic in the area since traffic dictates the frequency of the maintenance operations to be performed.

A second factor in evaluating maintenance is the cost of equipment to perform the maintenance, including repair costs. For example, floor maintenance equipment would include such items as scrubbers, vacuums, and mop buckets.

A third factor that should be considered in maintenance is the cost of expendable materials and supplies. Using carpet maintenance as an example, this category would include such items as shampoo, dry powder, paper rags, and spot removers.

USING PUBLISHED DATA

Data concerning comparative costs of various finish items is published frequently. Sometimes the findings of various studies are in conflict. In using published data three factors should definitely be considered. The first factor is the recentness of the information. Technological changes in materials and cleaning equipment occur rapidly. This can cause an analysis to be obsolete since it was not made using current materials.

The second factor is the source of the analysis. If the study was conducted under the auspices of manufacturer or trade association, it is probably best to take the findings with a grain of salt. On the other hand, if the study was made by a business school, the chances are that the data is less biased.

The third factor to be considered in using published data is consistency in the findings. If several recent studies arrive at approximately the same conclusions, then you can be more comfortable in applying the data to your selection process for finish items.

SECTION II—ORGANIZING

Food for thought:

Who's Steering?

Once there was a mighty oak tree which stood on the bank by a bend in the river. One day it began to rain. It rained for forty days and forty nights. During this time the river swelled and washed away the bank where the tree stood. On the fortieth day the mighty oak fell. It crashed into the river and rushed downstream in the swirling waters. On the tree were ten thousand ants. They were not organized to meet this new experience. Each ant thought that he was steering!

POINTS TO CONSIDER

Personnel in a maintenance department don't have to be like the ants on the log. They can be organized to perform their assigned tasks successfully.

This section takes a look at some of the facets of the management function of organizing. Chapter 9 deals with organizing your maintenance department. The other four chapters cover contracting, which should be recognized as a viable, cost-effective means of performing the mission of a maintenance department in many instances.

9 Organizing the Maintenance Department from a Bottom-line Point of View

The management function of organizing is the development of the organization structure and authority relationships required to achieve selected objectives. For a maintenance department organizing is the grouping of activities necessary to achieve the mission of the department and the assignment of each group to a supervisor.

In essence, organizing is the creation and maintenance of a structure of roles for employees in the maintenance department and contractors used by it. Some academicians may not agree that contractors should be included within the scope of organizing. But for many maintenance departments contractors are an integral part of accomplishing their mission, so let's recognize them as a part of the action.

Proper performance of organizing can lead to substantial cost savings. These savings can be achieved in labor, material, and capital investment in maintenance shop equipment.

- Labor savings can be achieved by reduction in headcount, avoidance of premium pay for overtime, reduction in shift differential pay, and efficiencies in shop equipment and material procurement.
- Material costs can be reduced by minimizing supply stock balances and duplicate tool crib items.
- Capital investment in shop equipment can be reduced by avoiding duplication of items.

Each maintenance department should tailor its organization to best meet its own particular mission. There is no one best way to organize, but there are three basic tasks which are necessary for efficiencies in cost and performance. These are:

1. Determining what work has to be performed (i.e., specifically defining the department's mission and the emphasis to be placed on the various types of work)
2. Deciding what work belongs together (i.e., the grouping of the functions to be performed)
3. Determining when the work is best accomplished (i.e., assigning operations by shift)

The accomplishment of these three tasks should not be a one-shot effort. A maintenance manager should be prepared to perform them periodically since conditions do not remain constant.

USING INDIVIDUAL ASSIGNMENT

In a small maintenance operation the individual assignment method is normally the most efficient. The maintenance manager personally directs the individual craftsmen who are assigned on a day-to-day basis to accomplish work. The organization is completely unstructured until the manager gets a Girl Friday. The Girl Friday begins to pick up some administrative functions such as timekeeping, compiling reports and ordering some parts.

Eventually as the department increases in size, the maintenance manager is no longer effective in personally directing all of the craftsmen in the operations to be performed. If the department begins to operate on more than one shift, the feasibility of personally directing all of the craftsmen rapidly declines. There are too many details and too many decisions. The maintenance manager who insists upon trying to do everything will find that he can no longer concentrate on the matters at hand, let alone do any planning with respect to organizing, staffing, controlling or directing. It is time for an organizational structure.

STRUCTURED ORGANIZATIONS

When a structured organization is created, it consists of a grouping of coordinated sections, each managed by a foreman or supervisor who acts with the authority of the maintenance manager on certain delegated functions. For a maintenance department there are three basic groupings for the line operations. These are:

- Operation
- Zone
- Shop

ORGANIZING BY OPERATION

In its purest form, organization by operation would consist of a separate group for inspection, maintenance, repair, overhaul, construction, housekeeping, and salvage. Each of these seven groups would have a foreman or supervisor responsible for the following activities.

Inspection would involve periodic inspection of equipment to insure safe and proper operation, assuring that periodic maintenance is performed, control of the quality of work accomplished by maintenance craftsmen, inspection of materials and parts received from vendors, and examination of items removed during repair or overhaul operations to determine the feasibility of repair.

Maintenance would consist of the lubricating, adjusting, and routine replacement usually classified as preventive maintenance. It would also include work necessary to restore equipment to operation on a quick-fix basis in the event of a breakdown.

Repair would consist of replacing parts to restore a piece of equipment to full operating condition and to alleviate undesirable conditions found during periodic maintenance or breakdown.

Overhaul would involve the reconditioning of equipment: teardown, replacement, reassembly, and testing.

Construction and rehabilitation would consist of building, modifying, and restoring structures.

Housekeeping would cover those janitorial and groundskeeping activities necessary to provide clean and orderly facilities and grounds.

Salvage would involve the reclamation and disposition of surplus material and scrap. The activity would not necessarily be limited to maintenance parts and materials. The handling of production scrap is often assigned to a maintenance department, since it is related to the clean-up and housekeeping activities. Collection and disposition of all surplus materials, equipment, and supplies is also often assigned to a maintenance department, since these operations are related to the maintenance inspection activity.

Transport maintenance generally follows the organization by operation approach since it is primarily concerned with a specific type of equipment (e.g., trucks or airplanes). The maintenance department is organized to some degree on the basis of the operations necessary to maintain this equipment. Figure 9-1 is an example of an organization by operation for aircraft transport maintenance.

Note that the operations of overhaul and inspection have maintained their identity. The salvage operation has been placed under administration, since it is closely related to material control and the stockroom in transport maintenance. The maintenance operation is placed under the service foreman. Because housekeeping and construction of the facilities are a minor part of the total mission, they have been assigned to the service section along with maintenance of the facilities.

Thus, as indicated in Figure 9-1, maintenance departments whose primary mission is transport maintenance generally use the operational approach to organization, but do not accomplish the groupings on purely an operational basis.

A department whose primary mission is facilities maintenance deals with many kinds of equipment. This type of maintenance department is seldom organized on an operational basis, because such an organization would result in inefficient utilization of manpower. Craftsmen assigned to the overhaul group could be sitting idle while the maintenance group was swamped with trouble calls. In most facilities' maintenance departments all seven operations are performed, but various facets of each operation

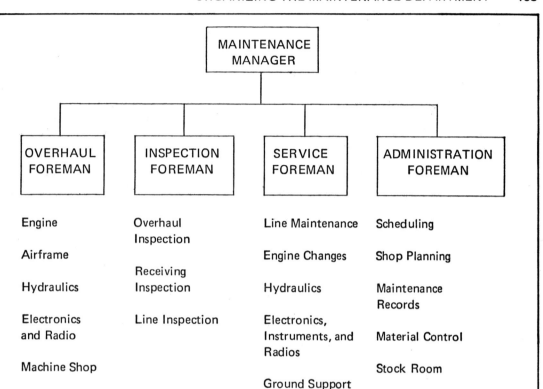

FIGURE 9-1

SAMPLE ORGANIZATION—AIRLINE TRANSPORT MAINTENANCE

are assigned to different groups. For example, the salvage activity of separating and collecting production scrap is combined with housekeeping in a manufacturing plant, while equipment salvage is assigned to the craft shops. Historically, maintenance departments with the primary mission of maintaining facilities have followed the zone or shop approach to organization.

ORGANIZING BY ZONE

A maintenance department organized on a zone basis assigns each maintenance group to a specific geographical location or area. The organization is geographically decentralized, with a foreman or supervisor over each group.

The advantages of zone maintenance are usually considered to be:

- Reduced travel time to and from jobs.
- More intimate knowledge of the equipment through repetitive experience with it.
- Improved job performance, due to greater interest resulting from a closer alliance with the objectives of a smaller group.
- More familiarity with the specific needs in the area serviced with improved relation between the users of the facilities and the maintenance department.

The main disadvantage to purely zone maintenance is the potential for inefficiency. There can be poor utilization of the labor force. In one zone the craftsmen can be up to their ears in trouble calls, while in another zone things are quiet and the men are performing low-priority work. Duplication of tool crib items and shop equipment with a low utilization rate can also result, since each zone group will endeavor to be self-sufficient.

ORGANIZING BY SHOP

This type of organization is essentially geographically centralized. The grouping is by craft shop. Each shop has a foreman responsible for all work done by his people. A shop may consist of one or more crafts. For example, see Figure 9-2. Here the electrical craft has its own shop, but sheet metal and air conditioning mechanics are in the same shop as the plumbers.

The advantages of a centralized shop organization are generally considered to be:

- Easier dispatching with more specialization by craft skill within the maintenance department.
- Higher-quality equipment justified for use in larger central shops.
- Better interlocking of craft effort through central control.
- More specialized supervision.
- Improved training facilities.
- Easier definition and administration of a system of plant-wide priorities making sure that available manpower is being applied to the most important jobs first.

There are disadvantages to a strictly centralized shop organization. First, in a large complex of facilities there is an increase in travel time. Second, a central shop organization does not facilitate familiarity with the needs of a given area or the intimate knowledge of equipment necessary in some maintenance operations.

CENTRALIZED VERSUS ZONE MAINTENANCE

The choice of centralized or zone maintenance was once a highly controversial subject. It is less so today, because most people who know anything about organizing a maintenance department have learned to take the best from each approach, based upon the type of facilities which are to be maintained. Here are some examples:

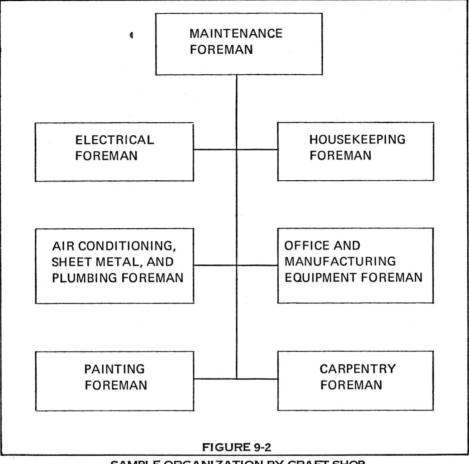

FIGURE 9-2

SAMPLE ORGANIZATION BY CRAFT SHOP

- A site in Southern California, consisting of 18 buildings of manufacturing, laboratory, and office area, assigns some crafts on a zone basis while other crafts operate out of the central maintenance shop. Carpenters, sheet metal mechanics, painters, and plumbers are centralized. Air conditioning mechanics, machine tool maintenance mechanics, landscape groundsmen, and janitorial crews are on area assignments.
- One large West Coast refinery uses essentially a centralized approach. Turn-around maintenance and major jobs are done by a task force which is assembled centrally and dispatched to the job. Shop work is also centralized, including major overhauls to equipment, fabrication, and small modification and construction jobs. Only routine maintenance, such as pump repacking, is handled on an area basis.
- A chemical plant in Texas operates almost exclusively out of seven area shops. There is also a central shop for instrumentation and electrical maintenance. The area shops provide zone maintenance, while the central shop has the high-quality equipment and specialized personnel for electrical and instrument maintenance.

- A large amusement park essentially handles all of its maintenance from centralized craft shops in which each craft has a foreman responsible for all work done by his people.

The layout and operating conditions of the facilities and the mission of each maintenance department should be continually analyzed to determine how much centralization or area maintenance is appropriate. Zone maintenance is appropriate when:

- Distances between the buildings are large
- Specific and intimate knowledge of equipment is essential to maintenance
- Idle maintenance manpower is less important than the value of the equipment being maintained (particularly production equipment)

A central shop organization is more appropriate when:

- Travel time to and from the job site is minimal or can be minimized by proper scheduling and planning
- The predelivery of materials and availability of required special tools can be readily accomplished
- Efficiency in maintenance manpower and shop equipment and better control of the accomplishment of the entire work load take precedence over the value of the equipment maintained

In production plants, the last consideration listed above is usually resisted by the manufacturing part of the organization. With centralization, production supervisors have less to say about setting priorities for their particular jobs. In shopping centers, office complexes, universities, hospitals, and the like, the argument for efficiency and better control through centralization is usually met with less resistance from the users of the facilities.

The use of radios has also increased the efficiency of personnel, particularly under a centralized system. There is immediate communication by the dispatcher or trouble-call desk operator to the craftsmen in the field. Two-way radios are generally used, although this equipment is more expensive than one-way devices. One site uses one-way radio to contact the craftsmen who then use a telephone to call back for instructions. Regardless of whether two-way or one-way radio contact is employed, the use of these devices greatly facilitates the handling of trouble calls and the dispatching of craftsmen in the field. The resulting increased efficiency will generally offset the equipment costs in a relatively short period of time.

LINE ORGANIZATION WITH STAFF UNITS

The arguments in favor of staff or supporting sections for a maintenance department are not new. The idea was advanced by the father of scientific management, Dr. Frederick W. Taylor, who proved it successful in production work improvement. The main reason for establishing support sections is to specialize activities so that they are accomplished more efficiently and the line personnel can devote their time to "doing."

ESTABLISHING A STOCKROOM OPERATION

Classifying Stock

Any maintenance department is required to keep a supply of materials and parts on hand if it is to operate efficiently. Items that are continuously used, such as standard nuts, bolts, screws, nails, and other hardware and high-usage materials (lubricants, paint, wood, wire, and so on), are generally referred to as *standard stock items.* For these items a minimum quantity is established and stock replenishment is initiated when the quantity on hand reaches the minimum quantity.

Items that are not part of the standard stock inventory are referred to as *special buy items.* These items are obtained on the basis of individual procurement requests, since there is no provision for a minimum quantity and automatic stock replenishment.

Low-value, high-usage standard stock items are sometimes designated as *open stock.* Craftsmen can obtain them directly from bins without the use of a stock requisition form. There is no control over who takes the stock or record of what job it was used on. Those maintenance managers who use an open-stock operation claim that using a stock clerk to issue the material and process the requisition paperwork costs more than the material being controlled. Further, if craftsmen are prone to pilfer, they will find a way to do it even in a closed stockroom operation. The one word of caution that open-stock advocates will generally express is to make sure that access to the open-stock area is limited to maintenance personnel and is not available to all employees of an organization. General access to an open-stock area invites a higher pilferage rate.

Setting Up Stockrooms

When stockroom operations require a separate service section, certain responsibilities should be clearly assigned to the group. Generally, these are as follows:

1. Issuing stock
2. Processing returned stock
3. Receiving standard stock and special buy materials and supplies from vendors
4. Stocking shelves and bins
5. Certifying receiving and invoicing documents
6. Conducting physical inventories
7. Reporting stock-outs
8. Initiating stock replenishment orders
9. Declassifying items as standard stock
10. Identifying and disposing of surplus
11. Identifying items that should be added to the standard stock inventory

The first ten items listed above should definitely be assigned to the stockroom operation. The activity of identifying items that should be added to the standard stock inventory may or may not be assigned to stockroom personnel. When the special buying operation is not a stockroom responsibility, it is more effective to assign this activity to personnel responsible for processing special buys. This is because repetitive special buys for an item is the best indication that the item should become part of the standard stock inventory. Since personnel in the special buying operations have the buy records, they are in a better position to identify repetitive buying action than stockroom personnel.

One maintenance department achieved a $25,000 savings by merely standardizing the inventory of lubricants used on machine tools.

Using Substores

When the maintenance department is organized on a zone or area basis, substore sections are frequently set up to support maintenance personnel in the given zones to eliminate travel time to and from the central stockroom and make standard stock items readily accessible to the craftsmen in the area.

A substore may be manned by a stores clerk or may be set up on an open-stock basis. If the substore operates on an open-stock basis, stockroom personnel should still be responsible for the following activities:

1. Stocking
2. Conducting physical inventories
3. Initiating stock replenishment
4. Identifying items which should be discontinued as substore items
5. Identifying new substore items

Some maintenance organizations make the mistake of assigning these activities to the craftsmen in the area. This is a mistake on two counts. First, craftsmen are usually interested in getting work done and are not prone to concern themselves about supply operations. Second, a craftsman receives substantially higher wages than a stock clerk. It is a misuse of his time to have him performing supply functions.

The one substore activity which craftsmen can help perform is the identification of new substore items. A conscientious craftsman who must repeatedly get the same item from central stores will generally request that it be placed in the substore he is using. At the same time, when central stores personnel or the substore clerk recognize that an item is being requested by a given zone on a repetitive basis, action should be taken to have it stocked in that zone's substore. Although craftsmen may help perform this activity, the basic responsibility for performance should be assigned to stockroom operations.

HAVING A PROCUREMENT SERVICES SECTION

Specializing the procurement function works hand-in-glove with the creation of a stockroom operation. The purpose of this specialization is to take the day-to-day ordering of supplies and requests for vendor services out of the hands of front-line foremen to free them to supervise the craftsmen.

The maintenance department in one automotive parts manufacturing plant in Michigan had around seventy craftsmen and had its foremen handling the procurement services function. As a result, the foremen spent more than half of each day in the office with telephones stuck to their ears. One clerk was hired to perform procurement services. This freed the five foremen to do their supervisory functions. Productivity increased around 20 percent, and the clerk's wages were paid for by the reduction in overtime pay.

A procurement services section normally is assigned the following activities:

1. Preparing purchase requisitions or purchase orders for special buy and standard stock materials, tool crib items, and capital equipment.
2. Coordinating deliveries with vendors.
3. Coordinating vendor services such as pest control, drapery cleaning, and carpet cleaning.
4. Locating sources of supply.
5. Coordinating the evaluation of new products.

Because of the close relationship with the stockroom operation regarding standard stock materials and supplies that normally constitute the largest portion of the procurement workload, this service section is frequently combined with stockroom operations. If they are not combined, then both supervisors should report to the same person to facilitate coordination between the two operations in fulfilling the overall supply function.

USING A TOOL CRIB

The purpose of a tool crib is to provide a storage area and an inventory control of portable tools and equipment that are not assigned to individual craftsmen on a permanent basis. A tool crib is normally a restricted access area with an attendant who signs out the items to individual craftsmen for use and subsequent return upon completion of the job.

The sign-out system should provide a record of who has the item, when it was signed out, and when it is planned to be returned. The craftsman is then accountable for the item until it is returned to the crib. If a signed-out item is needed on another job, the record makes it possible for that the item to be located for use.

The specific responsibilities which should be assigned to a tool crib operation are as follows:

1. Issuing items
2. Maintaining sign-out records
3. Receiving and inspecting returned tool crib items
4. Initiating action to service or repair tool crib items
5. Initiating action to retire or replace tool crib items
6. Obtaining additional tool crib items

Obtaining additional tool crib items involves two things. The additional tool crib item may be to supplement the existing inventory of an item. Procurement of additional like items may be warranted if there is not a sufficient number on hand to support the normal volume of requests for use. Tool crib personnel can generally initiate action for these duplicate items, since they are aware of usage. On the other hand, the addition may be an entirely new item for the tool crib inventory. In this instance the foreman should be responsible for identifying the need and the tool crib responsible for obtaining the item.

The types and quantities of items in a tool crib will vary based upon the mission of the maintenance department. In a maintenance department organized on essentially a zone basis, a tool crib operation may be established in each area. In addition, there may be a central crib to support plant-wide activities.

When tool crib operations are decentralized, there should be a consolidated inventory control system. Without such a system each tool crib will have a tendency to promote its own needs and refuse to cooperate in sharing its items even to meet emergencies, which can result in substantial duplication of items with an attendant low-usage rate and no loaning of an item if one crib's item is being repaired.

When one maintenance department installed a central inventory control system over its ten tool cribs, it was able to reduce its inventory costs by $15,000. This reduction was largely attributable to reallocation of existing items among the various cribs, installing a loaning method between the cribs, and, in some instances, providing supplementary items in a central crib to cover a repair cycle and emergencies.

Tool crib operations are often combined with stockroom operations, which permits cross-training and use of personnel in both operations, and provides economies of scale in handling the combined workload.

ACTIVATING A FINANCE STAFF

As soon as a maintenance clerk is assigned to keep track of costs, you have the beginning of a finance function. As organizations increase in size, their maintenance departments also become larger, and the expenditures for maintenance increase. Coordinating the planning of what these expenditures will be, finalizing departmental budgets, and keeping costs within the budget are of vital importance. If you are responsible for expending half a million dollars or more a year, that's a fair number of beans. You should have a bean counter to tell you where you are going to spend it and how fast you are spending it.

The usual responsibilities of a finance staff are:

1. Coordinating the preparation of departmental budgets and justifications
2. Analyzing cost performance against budgets
3. Investigating deviations from budgets to determine causes
4. Coordinating the preparation of budget revisions
5. Working with the accounting department

This last function is particularly important. Most accounting departments feel little concern about how well their chart of accounts and reports match the information needs of the maintenance department. A finance staff can perform an educational process and, hopefully, have the accounting department provide more meaningful cost information on maintenance operations.

USING A PERSONNEL SERVICES SECTION

Another specialization which evolves as a maintenance department grows is a personnel services section. The precise responsibilities assigned to this section or person will vary according to the nature of the organization that is supported by the maintenance department. A representative listing of some of the responsibilities is as follows:

1. Maintaining personnel record books, including posting vacation time, sick leave, overtime, accidents, illnesses, performance reviews, in-plant training, and outside education courses
2. Maintaining the maintenance department overtime records
3. Processing time cards and rate/shift change notices to payroll
4. Coordinating paycheck distribution
5. Coordinating employee service awards, educational reimbursement, and payroll deductions for savings bonds, the credit union, supplementary insurance, etc.
6. Assisting employees in filing medical insurance claims
7. Coordinating recruitment, including advertising, job notice posting, etc.
8. Coordinating job applicant interviews

As indicated above, the responsibilities cover a wide variety of activities, but are essentially oriented towards personnel administration activities, which are usually more efficiently handled by a specialized person or persons rather than by individual foremen. Even in smaller maintenance departments there is usually a Girl Friday who does some of these things on a part-time basis to relieve front-line supervision of these kinds of details.

PROVIDING A TRANSPORT SERVICES SECTION

In a large organization that has a substantial inventory of various types of vehicles (cars, trucks, personnel carriers, scrubbers, sweepers, fork lifts, mowers, and so on), the transport function is frequently specialized within the maintenance department. A transport services operation should be assigned responsibility for the following functions:

1. Ordering additional or replacement vehicles (leasing, buying or renting)
2. Servicing and repairing vehicles
3. Coordinating the retirement of vehicles
4. Fleet inventory management

The function of fleet inventory management is sometimes not assigned to the maintenance department or is not effectively administered. As a result, an organization generally spends more on its transport fleet than it should, because each department tends to feather its own nest rather than concern itself with the over-all costs to the organization.

One manufacturing plant in Los Angeles even reached the point of having separate motor scooters and fork lifts for each shift! Top management finally decided to have a fleet inventory management program administered by the maintenance department. Establishing a method of shared use of vehicles by various shifts and a centralized pool to provide spare vehicles reduced costs by over $150,000 per year.

The fleet inventory management function should not be limited to merely controlling the size of the inventory. It should also be responsible for evaluating what types of vehicles should be in the inventory. Standardization provides for interchangeable use of vehicles according to a repair cycle or on a pool basis.

In some instances, the size of the inventory of vehicles used by an organization, including the maintenance department vehicles, warrants having an in-house garage operation to perform most service and repair operations. To facilitate coordination, this garage operation should be part of the transport services section in the maintenance department.

It should also be noted that the standardization activity of fleet inventory management will reduce the number of spares required, as well as providing quantity discounts for this garage operation.

USING A WORK ORDER PROCESSING SECTION

The purpose of a work order processing section is to relieve line operations of the administration of paper work connected with doing the work. Some of the functions assigned to such a group are:

1. Preparation of work orders
2. Posting historical records on equipment items with cost and service data
3. Maintaining a status record on major jobs indicating where the job is and the percentage of completion of the jobs in design or in construction
4. Operating the trouble call and dispatching desk

The first three of the above functions are essentially clerical in nature. The last activity, handling trouble calls and dispatching, has to do with performing the management function of controlling. It is discussed further in chapter 18. When this last function is assigned to a work order processing section, what has really happened is the first step in creating a planning and control staff.

USING A PLANNING STAFF

The basic rationale in having a planning staff is that planning and "doing" require different skills, and a combination of these skills in one man is the exception rather than the rule. The craftsman who is "doing" should not have to interrupt his activity to plan the job sequence. Ideally, someone else should do the planning.

This same concept carries through to the separation of job performance from other functions, such as ordering and stocking supplies, maintaining administrative records, preparing departmental budgets, and administering contracts for purchased services. These functions are usually assigned to supporting groups without creating too much conflict. However, a good deal of conflict can arise when the planning function is assigned to a staff. This conflict usually results when the planning staff endeavors to issue instructions directly to the line supervision rather than through the individual who has line authority over the supervisors.

The functions assigned to a planning staff will vary. Some of those which are normally assigned are:

1. Work order processing
2. Scheduling work
3. Review of plans and specifications
4. Job planning
5. Applying job standards
6. Cost estimating (budgetary and planning
7. Coordination of construction by in-house crews
8. Construction supervision of contractors
9. Manpower planning

Properly utilized, a planning staff can be very efficient. Some sections, through job planning and the application of job standards to repetitive operations, have increased productivity of in-house maintenance crews in excess of 25 percent. As previously stated, if the maintenance department has around 200 craftsmen and over 100 active job orders involving construction, it is ripe for the use of a planning staff.

Having a Design Group

When a maintenance department is responsible for an extensive amount of modification and rehabilitation, it is frequently more efficient (in cost, time, and performance) to have an in-house design group rather than to farm out all the design work to an architectural and engineering firm. This design group should normally be capable of performing the following:

1. Conducting feasibility studies
2. Preparing architectural, electrical, and mechanical plans and specifications
3. Performing detailed cost estimating from plans and specifications
4. Preparing as-built drawings

It should be expected that the designs prepared by this group will not always be appreciated by the in-house construction crew who is going to do the work. There always seems to be a certain amount of conflict between planners and doers. Much of this conflict, however, can be resolved if:

1. The in-house construction crew supervision participates in the job walk and formulation of the design approach to accomplish the job.

2. There is a design review phase prior to release of the plans and specifications for construction.

USING A CONSTRUCTION COORDINATION SECTION

When a maintenance department is responsible for a great amount of construction work that is performed by contractors, a construction coordination section may be established to assure cooperation with the contractors.

The construction coordination section is normally assigned the following:

1. Participating in job walks with contractors bidding for the work and providing clarification of construction plans and specifications
2. Providing access clearances to contractors performing the work
3. Coordinating schedules
4. Administering the contract, including processing of field changes and contract modifications
5. Inspecting and making final acceptance of work performed

Inspectors assigned to the construction coordination section assure that the quality of the work is in conformance with the plans and specifications. They can achieve substantial cost savings in the area of field changes, as well as identifying schedule performance. Creating this specialized section is highly desirable when there is a continuous workload of contracted construction. One maintenance department estimates a net savings of $50,000 per year from its construction inspectors.

DETERMINING SHIFT COVERAGE

Requirements for shift coverage are essentially determined by following three factors:

1. The nature of the operations of the organization supported by the maintenance department
2. The type of work to be performed
3. Safety requirements

NATURE OF OPERATIONS SUPPORTED

The nature of the operations supported is often the primary factor in determining the number of shifts required for maintenance coverage, particularly for breakdowns.

In process industries in which the plants operate continuously, maintenance coverage is provided on a three-shift, seven-day-week basis in order to provide the service necessary for optimum production.

Where around-the-clock maintenance coverage is required, as much work as possible should be handled on the day shift. Lowest total maintenance costs and minimum downtime result from using the least coverage on the other shifts as can be tolerated from the standpoint of lost production time. This situation is attributable to the fact that suppliers, support services, and full supervision are available on the first shift. Further, the additional cost of shift differential pay is minimized.

THE TYPE OF WORK TO BE PERFORMED

The nature of the work often dictates the time or shift when it is performed. Examples are:

- Janitorial services in office areas are usually performed on a second shift when the office personnel are not occupying the area.
- Window washing on high-rise office buildings in some parts of the United States is done at night because during the day the water evaporates so rapidly that the cleaning cannot be done as effectively as during the night.
- Exterior painting of facilities is normally accomplished during daylight hours to avoid having to provide lighting for night operations.
- Parking lot sweeping is usually accomplished at times when the lots are least occupied.

SAFETY REQUIREMENTS

For some operations a breakdown may create a safety hazard so grave that maintenance coverage must be provided at all times. In other instances, ordinances require the presence of an operating engineer twenty-four hours a day, seven days a week. In either instance, the safety requirements dictate the extent of shift coverage required.

LOADING MAINTENANCE CREWS

The size of a crew on a given shift is primarily dependent upon the type and amount of work to be performed. The number of men on the shift should be balanced against the costs of holding men over on a premium time basis and the feasibility and cost of call-in personnel. To avoid overtime pay in providing coverage in seven-day-week operations, the techniques of relief men and odd-numbered manloading can be applied.

RELIEF MEN

Use of outside relief men can prove a cost-effective method of providing coverage on a seven-day-week basis. In essence, the relief man works two days per week as a fill-in on the days off of a regular full-time employee of the maintenance department. The savings accrue because the relief man is paid the normal shift rate instead of the overtime premium rate that would be paid to a full-time employee providing the coverage in addition to his normal work week.

DAYS WORKED	FIRST MAN	SECOND MAN	THIRD MAN	DAILY CREW SIZE
		THREE-MAN MAINTENANCE CREW		
Monday	X		X	2
Tuesday	X		X	2
Wednesday	X	X	X	3
Thursday	X	X		2
Friday	X	X		2
Saturday		X	X	2
Sunday		X	X	2

FIGURE 9-3

SAMPLE OF ODD-NUMBERED CREW LOADING

DAYS WORKED	MAN 1	MAN 2	MAN 3	MAN 4	MAN 5	MAN 6	MAN 7	DAILY CREW SIZE
			SEVEN-MAN MAINTENANCE CREW					
Monday	X		X		X	X	X	5
Tuesday	X		X	X	X	X		5
Wednesday	X	X		X	X	X		5
Thursday	X	X		X	X		X	5
Friday	X	X	X	X			X	5
Saturday		X	X	X		X	X	5
Sunday		X	X		X	X	X	5

FIGURE 9-4

SAMPLE OF THE RULE OF SEVEN

The use of relief men is normally restricted by two factors. The first is the availability of qualified personnel to fill the relief positions. The second factor is the permissability of the use of relief men under labor union contracts.

ODD-NUMBERED MANLOADING

The odd-numbered manloading technique is a method of avoiding the use of part-time relief men and overtime in providing seven-day coverage. Figure 9-3 shows an arrangement where a two-man crew is required. As indicated the first man works Monday through Friday, while the second man works Wednesday through Sunday. The third or odd-numbered man then works on those days when the first man or second man is off. On Wednesdays all three men are working. To avoid this overload in one day of the week, the rule of seven can be applied.

THE RULE OF SEVEN

When the crew is staffed by seven men, no overload takes place. As indicated in Figure 9-4, the daily crew size is consistently five men since the seven men rotate their days off. The rule of seven dictates that when alternate days off can be assigned to maintenance personnel for a seven-day-week operation, crew size in multiples of seven provides consistent headcount coverage.

The maintenance department at a large airline terminal reduced costs by $35,000 per year by applying the rule of seven.

10 Making the Most of Contractors in the Maintenance Function

No matter how diversified the in-house capability of a given maintenance department may be, situations will usually arise where economics or response time necessitate the use of outside firms for certain services or construction. Determining the extent of the need for contractors is part of performing the management function of organizing. This chapter presents some methods for using contractors effectively, both from the standpoint of cost and performance.

MAKING A POLICY

A policy with respect to the use of contractors is necessary as an aid in establishing the organization of the maintenance department. The policy should not only control what types of craft skills will be part of the in-house structure, it should also provide for efficient administration in the use of contractors.

EXAMINING PRACTICAL LIMITATIONS

In an analysis of the use of contractors, the first two practical factors to be considered are:

- The availability of adequate contractors
- Labor relations

If there are no contractors available in the geographical area, the only recourse is to import them. Importation is generally limited to large construction projects where the labor is brought in for a protracted period of time. Occasionally, for highly specialized work, a manufacturer's representative or a contractor may also be imported on a job basis. Generally, however, the nonavailability of contractors must be met by having a larger and more diversified in-house crew to perform the work expeditiously.

The other practical consideration of labor relations involves two possible factors. If the in-plant crew is unionized, the labor organization may be highly opposed to the prospect of any work being performed by members of another union or by nonunion workers. The reverse of this situation is where outside craft unions refuse to perform in-plant work, unless they are granted exclusive rights to all of the work or to clearly defined parts of the work. In either instance, severe limitations are placed on contracting.

CONSIDERING ECONOMICS

Criteria for division of work should be formulated through an analysis of the work assigned to the maintenance department and the evaluation of the relative costs of performance by in-house personnel or contractors.

Assuming that adequate contractors are available and that no appreciable labor relations constraints exist, the primary factor with respect to the use of outside contractors should be that of cost in relationship to:

- The type of work involved
- The amount of work involved
- The need date

The Type of Work Involved

The type of work dictates the degree of skill required. If the skill level is relatively low, sometimes the facilities and supervision of some other craft can be expanded to include the work. For example, maintenance carpenters often perform some of the work processes of a journeyman glazier such as work on locks and latches, door closers, hinges and panic devices, and some window glazing maintenance.

On the other hand, if the skill levels are relatively high or the equipment required is costly, exercising the option to contract is probably warranted. For example, most repair of computers or large electronic calculators is generally purchased from the manufacturer or his service representative because the skill level and test equipment required do not warrant having an in-house capability.

The Amount of Work Involved

In deciding whether to perform a task in-house, the amount of work is also a paramount consideration. One of the ultimate goals of organizing is to avoid idle time on the part of in-house labor. When there is insufficient work of a certain specialized nature to keep an employee occupied, it is probably best to contract the entire effort, particularly if the work also requires special equipment that would have a low utilization rate if the work were done in-house. Examples are parking lot sweeping or insecticide spraying in groundskeeping operations.

The Need Date

The third factor in determining the use of outside contractors involves two considerations. First, there may be an in-house capability, but the workload dictates that a contractor should be used in order to get the work done by the need date. Maintenance departments should not staff for the highest peaks. Ideally, contractors should be used as a reservoir to work off excessive backlogs.

Second, the in-house capability may be at less than optimum skill level and the additional costs in terms of downtime and quality of performance dictate that the job should be contracted. For example, in-house maintenance carpenters may be capable of installing dropped ceilings. However, a contractor who specializes in the work can probably do a large job faster and at less total cost, because his people are better trained to do the work expeditiously.

One maintenance department found that the dollar cost difference on dropped ceiling work in areas greater than 500 square feet averaged 15 percent higher when the work was done in-house. This difference was primarily attributable to a higher productivity rate on the part of the contractor's personnel who specialize in that type of work. Contracting the work saved over $23,000 per annum.

APPLYING COST FIGURES

In analyzing the cost of doing the work in-house, all maintenance cost factors should be considered. These are:

- Wages
- Related payroll expense (fringe benefits)
- Supervisory and administrative costs
- Equipment costs (expense and capitalized)

Where applicable, a cost factor for production downtime may also be added to the computation. Some maintenance departments also include an element for quality of performance. This element is somewhat nebulous, but what they are trying to do is measure and quantify in terms of cost the fact that the in-house crew may perform slower than a contractor specialized in performing the work. Examples are laying brick or concrete block, laying carpet or floor tile, stacking glazed tile in restrooms, and roofing or framing on building additions.

With respect to analyzing the contractor's price, it should be recognized that he has to add the element of profit to his costs for wages, fringe benefits, overhead, and equipment. This profit markup is normally added to his costs for both materials and labor. Travel time may also be another consideration. Contractor maintenance service personnel are generally paid while they are going to and from the job site, which can add a couple of hours of nonproductive time to the contractor's labor costs.

CLASSIFYING CONTRACTS BY THE EXTENT OF USE

The use of outside contractors has a marked impact upon how a maintenance department is organized. When contractors are used extensively the maintenance department structure should be designed to facilitate their use. In analyzing the use of contractors, it is sometimes beneficial to classify the types of contracting on the basis of the extent of use of the services.

ROUTINE CONTRACTING

Some contracts can be let for performance on a scheduled, routine basis. Examples are contracts for pest control, waste removal, escalator and elevator service, parking lot sweeping, exterior window washing, and janitorial services. For these types of services a prescribed schedule is established which the contractor is to follow in providing the services. When these types of scheduled services are used extensively, a purchased services group or unit is often created within the maintenance organization to provide the day-to-day supervision, which prevents the line foreman concerned with the supervision of in-house labor from getting involved in this kind of detail.

UNSCHEDULED INTERMITTENT CONTRACTING

When contractors are used on this basis it is generally to reduce peak workloads or to provide services that cannot be performed in-house. The use of the contractor is unscheduled in the sense that the services are not requested until the need arises. However, the use of contractors should not be approached willy-nilly so that a frantic search begins to find someone to do the work when the need arises. There should be an analysis made of potential contractor's available and a system devised on how to deal with them. In some instances a contract may be let for the services on an open-call basis.

Open-Call Contracting

An open-call contract provides for the performance of services on an as-requested basis. It eliminates the time and administrative effort required in letting a separate contract or purchase order each time the need for the services arises. Examples of this type of contracting are:

- Carpet or drape cleaning
- Carpet installation
- Window blind maintenance
- Purchased labor

Because a large amount of carpet or drape cleaning will be required during the year, bids are let and a contract awarded to an organization with the specialized equipment and trained personnel capable of doing this type of work. As the needs arise, the contractor is called in to do the work.

Where an extensive number of moves and modifications are involved, carpet installation contracts can be let on the same basis, either with the contractor providing the carpeting, the installation materials, and labor, or with the carpeting and materials being provided to him.

In office complexes with a large number of window blinds that require servicing and replacement of horizontal or vertical slats, an open-call contract for materials and labor can be used effectively to provide for maintenance. Rather than trying to stock the materials and provide in-house labor, better service can be provided at approximately the same cost by a vendor in metropolitan areas.

An open-call contract for purchased labor may be used as a means of supplementing in-house personnel during peak workloads. This type of contracting is discussed further in chapter 11 with respect to construction. It can also be used effectively in high-demand periods for turnaround maintenance in process production plants. Although the vendor coordination may be accomplished by a purchased services section, it should be recognized that supervision of purchased labor personnel, while actually on the site, should be done by the craft supervisors of the maintenance department. The work of purchased labor personnel should be integrated with the effort of in-house labor. This integration is best achieved by the supervision of the maintenance department responsible for supervising the in-house craftsmen. If the purchased services section of the maintenance department endeavors to supervise the purchased labor personnel, invariably the efforts will become disjointed.

Job Unit Contracting

When job unit contracting is used, a separate contract or purchase order is let for the work each time the need arises. This type of contracting is normally used for construction projects, which are discussed further in chapter 11. Other examples of job unit contracts are resurfacing of parking lots, exterior painting of buildings, and rehabilitation of machine tools.

QUALIFYING POTENTIAL CONTRACTORS

Regardless of whether a maintenance department contracts various types of work on a routine or intermittent basis, potential contractors should be checked out. A check can usually be accomplished by a few telephone calls or letters to the contractor's previous and current customers, banker, bonding agency, credit references, suppliers, and subcontractors.

The use of these types of sources should not be construed to mean that only large firms should be considered. Often, small local firms can perform work at considerably lower rates, because they do not have the overhead costs which larger companies have. Also, you may be a little customer to a big contractor, but a big customer to a little contractor. In the latter instance you are more apt to get better performance. The axiom "bigger ain't necessarily better" applies to using outside firms.

In evaluating a contractor the following criteria should be applied:

- Experience
- Capability
- Reputation
- Availability

EXPERIENCE

There is no substitute for experience. If a contractor cannot demonstrate that he has previously performed the type of work involved and cannot provide suitable references from satisfied customers, awarding him the work contains a definite element of risk. You may pay dearly for his learning. This risk should be analyzed from the standpoint of how critical things would be if he falls on his face. If he is new at the game, but it is a small job where defects are readily remedied if he fails to perform properly, it might be worth the risk. A contractor will often remember his first customers when he was just starting up and, out of appreciation, give them priority service over the years.

CAPABILITY FOR PERFORMANCE ON THE PROJECT

The term capability refers to the extent of the contractor's resources to do the specific project at hand. His resources consist of the necessary equipment, trained manpower, working capital, and technical and administrative know-how. For a general contractor on construction work, another vital factor that should be considered as a part of his capability is his team of subcontractors.

If a contractor cannot demonstrate his capability to perform the type and scope of work, there is a risk in doing business with him. Again, as with a contractor who cannot demonstrate his experience, you may be willing to take a risk if you can recover rapidly from a lack of performance.

REPUTATION

The term reputation refers to the axiom of not doing business with a guy who operates out of a phone booth. It is usually best to do business with an established firm that has a track record for honesty and integrity in dealing with its customers. To do otherwise may result in a contract upon which there is poor performance or even default. As with the factor of experience, if the risk is small, you may want to take a chance with a new firm because of the potential of getting better service in the future as a first customer.

AVAILABILITY

The criterion of availability pertains to the contractor's ability to perform the work within the required time frame.

Availability or response time should be a paramount consideration in evaluating a potential contractor. He may have a wealth of experience in the kind of work to be performed, thoroughly trained workmen, and a excellent reputation, but be too busy to provide the services in the period required.

One method of ascertaining this ability on construction work is to prescribe in the bid the time of start and completion of the work. For service contract bidding, a response time can be stipulated in terms of so many hours from receipt of a trouble call.

USING ARCHITECTURAL AND ENGINEERING FIRMS

Proper design and administration of extensive construction projects require a high technical capability level, particularly if the construction involves more than just brick and mortar. If there is extensive electrical and mechanical engineering required for the facilities, trained and licensed engineers are required in the preparation of the design and the review of proposed changes during the construction phase. Creating an in-house design group with this high level of capability is usually uneconomical. The services of an architectural and engineering (A&E) firm are used on extensive construction projects.

The services obtained from an A&E firm may comprise any or all of the following work elements:

- Design
- Construction supervision
- Preparation of as-built drawings

DESIGN

Design involves the preparation of plans and specifications and a cost estimate for the construction. It also consists of obtaining a plan check by local government agencies, which must approve the design for construction.

CONSTRUCTION SUPERVISION

Construction supervision consists of providing technical support during the selection of a construction contractor, administering the construction phase for compliance with the plans and specifications, and reviewing and authorizing field changes.

To completely delegate construction supervision to the A&E firm that prepared the design is inviting trouble. No matter how reputable the A&E firm is, the maintenance department is ultimately responsible for the successful completion of the project. Although representatives of the A&E firm may be present to provide clarification during job walks in the selection process, a representative of the maintenance department should also participate. Someone in the maintenance department should also keep tabs on what is going on during the actual construction. Otherwise, the deviations between the design and actual construction may be acceptable from an engineering point of view, but poor from the standpoint of having to maintain the facilities after they are constructed.

AS-BUILT DRAWINGS

Preparation of as-built drawings consists essentially of updating original design drawings to reflect field changes made during construction.

Normally, as-built drawings should be made by the same group that prepared the design because they are familiar with the drawings. If an A&E firm did the design, preparing as-built drawings in-house will usually take substantially longer, and the preparation of as-built drawings will cost more, even though in-house draftsmen cost less per hour than those of the A&E firm.

On major construction jobs, one facilities maintenance department discovered that as-built drawings prepared in-house cost them 12 percent more than as-built drawings done by the A&E firm. Also, the drawings were usually completed and available at least thirty days sooner when the A&E firm did the work.

LETTING WORK

When contracting on a job basis is done extensively, the preparation and maintenance of lists of qualified contractors should be an ongoing task. Need date and overall performance time is often the paramount consideration in deciding to contract work. To support the need date, no time should have to be spent shopping around for potential contractors. This spade work should have already been accomplished. A list of qualified contractors for the type of work involved should be readily available.

Figure 10-1 is a rather extensive listing of the various categories by which contractors are classified by a maintenance department responsible for a site consisting of twenty-one buildings and some 200 acres of landscape and parking area. With this data bank of qualified contractors, they are in a position to let a contract rapidly when the need arises.

In selecting contractors for quotations, one maintenance department looks at the size of the job and the trades involved as the key considerations. Their practice is to invite the previous two low bidders for a particular type of job and to add at least two additional bidders from a current list of prospective bidders. This standard practice rewards previous low bidders with the opportunity of bidding on new work, while the high bidders are replaced with other contractors interested in doing business.

In bidding for major modification work this same maintenance department requires that the general contractor's bid be broken down by subcontract. The breakdown is reviewed against in-house cost estimates. This method of review accomplishes the following:

1. Identifies which subcontractors have the greatest influence on total price
2. Validates in-house cost estimates
3. Updates current in-house pricing data
4. Provides feedback on labor rates for the various craft skills required for the job
5. Indicates items of work which were missed by the contractor which must be considered prior to contract award to assure that the entire scope of work is covered

FILE CATEGORIES OF POTENTIAL CONTRACTORS

General Construction

Electrical

Mechanical (heating, ventilating, and air conditioning)
Insulation
Plumbing
Window Glazing and Replacement
Roofing
Partitions
Cabinet Work
Painting
Paving and Parking Lot Services

Concrete Work

Crane and Hoist Installation and Repair

Demolition

Fence Installation

Fire Sprinkler Installation

Floor Covering
Carpet Cleaning
Drapery Cleaning and Installation
Wall Carpeting and Papering
Sign Installation and Maintenance

Furniture Repair

Landscape Gardening
Weed Abatement

Automatic Parking Systems

Window Washing and General Building Maintenance

FIGURE 10-1

SAMPLE BREAKDOWN OF POTENTIAL CONTRACTORS

SIPHONING OFF CONSTRUCTION WORK

Even when there is an in-house construction crew, the use of contractors should not be ruled out. If there are marked peaks and valleys in the workload, the in-house force should be staffed so that it can handle a workload slightly above the valleys. This action should be taken in anticipation of being able to defer some of the work in the peak periods. Contractors should then be used to work off excessive backlogs in the peak periods.

One maintenance department located in the Los Angeles area, where qualified contractors are readily available, evaluates each individual construction project. To reach a decision on in-house versus contracted construction, the following factors are considered:

1. Priority of project in relationship to other projects
2. Overall cost
3. Need date
4. Timing (i.e., shift on which the work must be performed)
5. Security of work area (risk of product disclosure or process know-how)
6. Availability of in-house personnel and skills to meet need date
7. Key materials and equipment required

In order to establish the proper design approach, the decision to contract is made before the project is released for design. The decision is made early because the extent of plans and specifications required for use by the in-house crew require 30 to 40 percent less detail than when the job is to be done by outside contractors. When the in-house crew is assigned the work, design time is reduced, which results in a savings of some $25,000 per year in design labor.

ORGANIZING TO USE CONTRACTORS

As previously stated, the maintenance department should be organized in a manner to facilitate the use of contractors. There is no one best way to do this, but responsibilities for certain functions should be clearly assigned. Figure 10-2 indicates sixteen major functions that should be accomplished and how they are generally assigned in a maintenance department consisting of a line organization with staff units. Depending upon the organization, the maintenance department may have to go through the purchasing department in the contracting process, which is why some functions are labeled "Purchasing Department/Purchased Services Group." Although the actual contracts may be let by the purchasing department, responsibility for day-to-day administrative activities should be clearly assigned to specific personnel within the maintenance department.

CONTRACTING OPERATIONS

FUNCTION	GROUP WITH PRIMARY RESPONSIBILITY	GROUP PROVIDING SUPPORT
1. Qualifying potential contractors	Purchasing Dept./Purchased Services Group	
2. Maintaining lists of potential contractors	Purchasing Dept./Purchased Services Group	Construction Coordination Group
3. Negotiating sole source construction contracts	Construction Coordination Group	Purchasing Dept./Purchased Services Group
4. Issuing request for quotation	Purchasing Dept./Purchased Services Group	
5. Conducting job walks and bidder's conferences	Construction Coordination Group	Purchasing Dept./Purchased Services Group
6. Conducting construction contract bid openings	Purchasing Dept./Purchased Services Group	
7. Issuing contract award notices and notices to unsuccessful bidders	Purchasing Dept./Purchased Services Group	
8. Issuing contracts and amendments	Purchasing Dept./Purchased Services Group	
9. Supervising construction contracts	Construction Coordination Group	
10. Verifying construction contract invoices for payment	Construction Coordination Group	
11. Administering warranties and guarantees	Purchasing Dept./Purchased Services Group	Construction Coordination Group and Craft Foremen
12. Reviewing construction plans and speci-fications prepared by A & E firms	Construction Coordination Group	Design Group
13. Reviewing construction cost estimates prepared by A & E firms	Design Group-Cost Estimating Section	Construction Coordination Group
14. On-site supervision of craft purchased labor and verifying invoices	Craft Foremen	
15. On-site supervision of design purchased labor and verifying invoices	Design Supervisor	
16. Providing liason for service contracts and open-call contractors	Purchasing Dept./Purchased Services Group	Craft Foremen

FIGURE 10-2

ORGANIZING TO USE CONTRACTORS

Note that all sixteen functions are not the primary responsibility of the same section within the maintenance department. A purchased services or contracting group may be established to perform most of them, but others are assigned to a design or construction coordination group, because of the engineering knowledge required to properly perform the function. Note also that although one group may have primary responsibility, other groups are to provide assistance in the performance of some of the functions as indicated in the "GROUP PROVIDING SUPPORT" section of the figure.

11 Administering Contracts for Maintenance Construction Projects

Contracts for construction projects can take many different forms. A maintenance manager who administers construction contracts should be fully aware of the advantages and disadvantages of the various types of contracts that are available for his use. Otherwise, he stands a chance of paying more for the work than he should have, losing time in properly completing the job, and possible litigation.

Six types of contracts are presented in this chapter, with some guidelines on deciding when to use each type. The last two sections of the chapter cover the need for the use of provisioning contract clauses and the vital role of a construction supervisor.

FIXED-PRICE CONTRACTS

Fixed-price contracts are also referred to as stipulated-sum or lump-sum contracts. Regardless of which term is used, under this type of contract the work is done for a set price, based upon competitive bidding or negotiation.

BENEFITS

The main advantages to fixed-price contracting are three. First, there is advance knowledge of what the construction cost will be. Second, there is an undivided responsibility for the project, which rests with the single contractor. Third, the contract is flexible with respect to changes.

DRAWBACKS

There are disadvantages to a fixed-price contract. An unscrupulous contractor may attempt to cut quality to obtain maximum profit or an incompetent contractor may be awarded the contract on the basis of lowest bid. Extra work and changes may be overpriced by an unscrupulous contractor. He has the advantage of already being on the job site and knows that there is little likelihood that the extra work or changes will be awarded to another contractor. Also, special care may be required to obtain complete contract compliance.

COST-PLUS-FIXED-FEE CONTRACTS

Under a cost-plus-fixed-fee contract, the contractor is reimbursed for the cost of all work performed and is paid a fixed sum or fee for his services.

BENEFITS

There are three major advantages of this type of contract. First, the total time for the project may be reduced by concurrent construction and completion of drawings and specifications. Second, the contractor's incentive to cut quality is removed. Third, changes in work are readily made.

DRAWBACKS

The major disadvantage to this type of contract is that the total cost of the project cannot be determined in advance. Also, extensive changes in work may require changes in the fee. Keeping costs low is largely dependent upon the integrity of the contractor.

COST-PLUS-PERCENTAGE-FEE CONTRACTS

Under this type of contract, the contractor is reimbursed for the cost of all work performed and is paid a fee based upon the final cost of the work.

BENEFITS

The major advantage to this type of contract is that changes in work are readily made, since the fee is not fixed. This type of contract also has the advantage of reducing project time by concurrent construction and completion of drawings and specifications.

DRAWBACKS

A marked disadvantage is that the contractor's self-interest in high costs is strong, since his fee is based upon total costs. Also, there is no advance knowledge as to what the total costs will be.

GUARANTEED-MAXIMUM CONTRACTS

A guaranteed-maximum contract is another variation of a cost-plus contract. It is sometimes referred to as a cost-plus-fixed-fee with guaranteed cost. Under this type of contract, the contractor is reimbursed the cost of all work performed and is paid a fixed fee for his services, but the contractor guarantees a maximum cost figure. The owner and the contractor may share in any savings below this guaranteed maximum cost figure.

BENEFITS

The main advantage of this type of contract is that a maximum cost is assured. If the contractor shares in the savings below the guaranteed maximum, he has an incentive to reduce costs.

DRAWBACKS

The disadvantage of a guaranteed-maximum contract is that drawings and specifications must be virtually completed before a contractor is normally willing to establish a guaranteed price. Thus, savings in construction time by concurrent completion of drawings and specifications is reduced. Also, unless savings are shared, keeping costs low is dependent upon the integrity of the contractor.

PURCHASED-LABOR CONTRACTS

A purchased-labor contract is one whereby the seller provides personnel to the buyer at specified hourly rates. The contract contains an hourly rate schedule for each of the various types of personnel or craftsmen that may be obtained. The contract is usually an open-call type, whereby the buyer may requisition the amount of personnel in the various craft categories on a daily basis as they may be required during the life of the contract.

BENEFITS

One advantage of this type of contract is flexibility. The buyer may readily supplement his own, in-house construction crew without adding full-time personnel to his payroll. It provides the buyer flexibility in rapidly obtaining additional personnel during peak workload periods or the vacation periods of his own personnel.

One maintenance department in Southern California also found that it was the least expensive way of performing minor construction and rearrangement jobs. Purchased labor was less costly than the total cost for a comparable full-time employee on their own payroll. When the related payroll expenses of the company's employee benefits program were added to the actual wages paid an employee, the sum was greater than the rates charged by the purchased labor contractor.

DRAWBACKS

One disadvantage of a purchased-labor contract is control. The buyer must accept the personnel provided by the seller. If they are not qualified craftsmen, fail to properly follow instructions, or goof off when they should be working, the buyer has little recourse. He may gripe, but he generally pays the bill. Thereafter, he just doesn't ask for that type of labor again or changes to another supplier. As an added feature to the contract, the buyer may have the right to expressly exclude certain personnel, by name, from being assigned to him, which provides subsequent control after non-achievers have made themselves known. The buyer, however, does pay for something which he does not get while the nonachievers identify themselves.

The use of a purchased-labor contract also may be fraught with labor union problems, particularly if the purchased laborers are not unionized. The main grievances usually center on restricting overtime or using nonunion personnel. Problems may also arise if a contractor on the same job site employs unionized personnel. The labor union situation should be analyzed thoroughly before purchased labor is brought onto a job site. Otherwise, you may end up creating more problems than you have solved.

TIME-AND-MATERIAL CONTRACTS

A time-and-material contract is another variation of cost-type contracting. The contractor is reimbursed for the cost of all materials and labor used on the job. For all labor expended, he is reimbursed at stipulated rates for each of the various types of personnel, including supervision. These rates include a markup on the actual wages paid to cover the contractor's overhead and profit. The materials provided are normally invoiced at cost, plus a handling charge. This handling charge is usually a percentage markup over cost whereby the contractor covers procurement overhead and gets some profit.

BENEFITS

The main advantage to this type of contract is flexibility in changing the scope of work. The contractor merely provides whatever materials and labor are required for the project at the agreed-upon rates. There is no additional negotiation required with respect to fee, since the fee is included in the rates. Another advantage may be a reduction in time of completion of the project by concurrent construction and completion of plans and specifications.

DRAWBACKS

The major disadvantage to a time-and-material contract is that the total cost of the contract cannot be determined in advance. There is also a lack of control over the costs. The contractor has no incentive to reduce labor, since by doing so he would reduce his profit. He also has no incentive to avoid wasting materials.

Under a time and material contract, supervision of the personnel is normally provided by the contractor and there is little opportunity for the buyer to exercise control over the quality of the personnel used on the job.

Because of these factors, some maintenance managers regard a time-and-material contract as a license to steal, even if the most reputable contractors are used.

A time-and-material contract may also be fraught with the same labor relations problems as a purchased-labor contract. Therefore, the labor union situation should be analyzed thoroughly before you contract on a time-and-material basis.

DECIDING WHAT TYPE OF CONTRACT TO USE

In the previous sections of this chapter six types of contracts have been described with some of their advantages and disadvantages. Which one should be used is largely a matter of trade-off between time, cost, and the scope of the project.

MINOR CONSTRUCTION JOBS

Time-and-material contracts should be used rarely, because of the major drawback with respect to controlling costs. Application is normally restricted to specialty-type work required as only a part of a minor project such as dropped ceilings, office partitioning, and sheet metal work. In theory, a contractor can do the work more efficiently than an in-house maintenance crew, because his personnel are specialized in the type of work and can, therefore, perform it faster. In using a time-and-material contract for specialized work, one must first insure that no labor conflicts will arise from the presence of the contractor's personnel and the in-house maintenance crew on the same job site.

When *time* is the paramount consideration on a minor construction job, such as an area rearrangement, a purchased-labor contract is often used. Assuming there are no labor relations conflicts, a maintenance manager merely supplements his in-house construction crew with the numbers and types of craftsmen he needs to rapidly complete the job. Since the purchased-labor craftsmen constitute a small portion of the total crew, they are supervised by the in-house maintenance foremen and leadmen assigned to the job. Slackers should be quickly identified. At the end of a shift they should be immediately reported to the purchased-labor supplier and replacements requested.

One facilities maintenance department uses purchased-labor contracting extensively during peak workloads for minor modification work. The department finds it to be the quickest way to respond in getting the work done in comparison to other methods of contracting. In addition to being able to rapidly expand and contract the work force, they also estimate that they save some $45,000 in payroll and related payroll expenses per year.

When there is no in-house construction crew available to do the work with supplementary purchased labor, or the use of purchased labor is not readily feasible, a fixed-price contract is generally used with competitive bidding. The scope of work is usually readily identifiable and does not entail many contingencies, since drawings and specifications can be completed prior to the bidding. Thus, there is no need for a cost-plus contract, under which costs are not as readily controllable.

The full benefits of fixed-price contracting are most readily achievable on minor construction projects. Although comparative cost data is not easily calculated, one maintenance department estimates that it generally saves 10 to 15 percent on minor construction jobs by using fixed-price in lieu of cost-plus contracting. This saves them from $25,000 to $40,00 per year.

MAJOR CONSTRUCTION JOBS

On major construction projects such as a large addition to a building, the complete rehabilitation of a facility, or the rearrangement of a substantial portion of a building, a fixed-price or cost-plus contract is usually used.

When *cost* is the predominant factor on a major construction project, a fixed-price contract with competitive bidding is normally used. However, the quality of the work should be assured by having precise specifications and adequate construction supervision. Changes should be minimized by having a thorough review to finalize the design prior to bidding the work.

When *time* is the paramount consideration on a major construction job, a cost-plus type contract is often used. Of the three types of cost-plus contracts, a cost-plus-percentage-fee contract is not generally used, unless substantial changes in the work are anticipated which would require many changes to a fixed-fee arrangement. There are better chances for lower costs under a cost-plus-fixed-fee contract than a percentage-fee contract. The savings in total project time are the same for both types.

A guaranteed-maximum contract does not normally achieve the time savings which are obtained from a cost-plus-fixed-fee or percentage-fee contract because a contractor will not provide a maximum-cost figure until the design is virtually completed. A guaranteed-maximum contract is not the best type of contract by which to reduce total project time. However, this type of contract has proven highly effective both in terms of time and cost where an owner, architect, and contractor have a long-standing working relationship in a site development. In this case the contractor is familiar with the architecture and is willing to provide a guaranteed cost long before the design is completed. A maintenance manager, however, is seldom in this type of situation. Therefore, a guaranteed-maximum contract is not used by most maintenance departments.

MIXING CONTRACT TYPES WITH THE SAME CONTRACTOR

Another consideration in selecting the type of contract to be used is whether the contractor is already performing other work on the site. If he is, this should be brought into consideration in contracting additional work with him.

For example, assume that the contractor is performing construction work as a general or subcontractor under a fixed-price contract. Included in the fixed price is the use of certain equipment and supervisory costs. If the same contractor is now given a cost-plus contract for concurrent work, there is the opportunity for these costs to be billed against the cost-plus contract. As he endeavors to maximize his profits, you could end up paying twice for the same equipment and supervisory personnel. To guard against this possibility, you may not want to give a cost-plus contract to this same contractor. If you do give one to him, there should be close construction supervision and a careful audit of invoices on the cost-plus contract to assure that the changes are legitimate.

One maintenance department discovered that a contractor had invoiced some $25,000 in charges to a second, cost-plus contract, which were probably allocable to his fixed-price contract. However, close construction supervision had not been exercised and the records were not precise enough to prove that the contractor had actually misapplied the costs. Since there was little chance of winning a case in a court of law, the contractor was reimbursed. It was a stiff price to pay for learning how to properly contract for construction work.

APPLYING PROVISIONING CONTRACT CLAUSES

The purpose of provisioning contract clauses is to assure that a general contractor and his subcontractors are contractually obligated to provide certain information. Generally this data should consist of the following:

- Marked-up drawings showing all details of construction as actually performed that differ from details shown on the contract drawings, which are used to prepare as-built drawings
- Manufacturer's operating manuals, warranties, spare parts listings, and related data for building equipment installed during construction
- The names of the manufacturer's representatives closest to the construction site who can provide service and technical information on the operation of the building equipment

On major construction, rehabilitation or new construction projects, there is a definite need for provisioning contract clauses. When a maintenance department assumes responsibility for the facilities without having this data, a lot of work has to be done to get the information necessary to adequately plan and perform the maintenance program.

To provide some lead time for maintenance planning and the procurement of spare and replacement parts, the data on building equipment should be required to be delivered prior to final acceptance of the work. On major new construction projects, some contracts establish the delivery date for this data as 120 days prior to the scheduled final acceptance date. Others use a ninety-, sixty-, or forty-five-day delivery date prior to final acceptance. Usually, a contractor is tardy in getting all of the data delivered, so a thirty-day lead time should be added to whatever date you really need the data in order to start the detailed planning of your maintenance program.

Some maintenance departments fail to do any maintenance planning or perform any maintenance while the facility, including the building equipment, are under warranty because they are under the naive assumption that the general contractor or equipment manufacturer will make good on any defects. Normally, there is a warranty period during which the contractor is responsible for any defects. However, if the owner of the facility has failed to perform prescribed maintenance, both the general contractor and the equipment manufacturer are off the hook.

One food processing plant had to pay $8000 for a new refrigeration compressor within four months of occupying their new plant. Neither the general contractor nor the equipment manufacturer would assume any responsibility for failure of the original compressor because there was no evidence that any of the prescribed maintenance had been performed.

At another plant, $5800 had to be paid for a new compressor. In this instance maintenance had been performed, but the wrong lubricants had been used, so the equipment manufacturer was absolved of his warranty provisions.

USING A CONSTRUCTION SUPERVISOR

When a decision has been made to contract for construction, a construction supervisor should be immediately assigned to the project. Normally, the decision to contract should be made prior to the start of design so that the construction supervisor can actually participate in the design phase and become thoroughly familiar with the nature of the project. He is then ready to perform the following tasks:

1. Participate in job walks and provide clarification on the plans and specifications to potential contractors
2. Review the contract plans and specifications with the contractor awarded the work prior to the start of construction
3. Act as the sole point of contact for the contractor in dealing with the maintenance department
4. Coordinate the access of contractor personnel to the job site
5. Provide general contract administration including construction permits, quotations for changes in the scope of work, authorizing field changes, and processing contract change and amendment documents
6. Receive and review marked-up drawings reflecting changes from original design drawings for preparation of as-built drawings
7. Receive manufacturer's manuals and related data on building equipment
8. Inspect, prepare punch lists on defects, and make final acceptance of the work

Note that tasks six and seven above cover items in provisioning clauses.

The construction supervisor can have many different job titles, such as construction coordinator, construction liaison man, and major construction representative.

Regardless of what he is called, to properly accomplish all of the tasks listed above, you need a person who is thoroughly versed in the type of construction work called for by the project. There is no place for a neophyte, and the need for a trained and experienced construction coordinator cannot be overemphasized.

Figure 11-1 is a sample of a form that is used by a construction coordinator to certify completion of a contract. Figure 11-2 is the second page of the form, which is prepared by the construction coordinator for performance evaluation. Note that the form covers both the general contractor and the subcontractors used on the contract. Figure 11-2 provides a means for using the construction coordinator as a resource in evaluating contractors.

CERTIFICATION OF COMPLETION
OF MAJOR CONSTRUCTION CONTRACT
AND CONTRACTOR EVALUATION SHEET

Contract No. _____ Name of Contractor _____

The undersigned hereby certifies that the following items have been accomplished and the final payment to the Contractor is hereby approved.

1. All work to be performed under the terms of the contract has been completed including all items noted on "punch lists" citing discrepancies in the construction which were to be corrected prior to final acceptance.

2. Manufacturer's manuals, warranties, and related data for building equipment installed during construction have been provided by the Contractor in accordance with applicable contract provisions. Such data was provided to Maintenance for establishment of preventive maintenance records _____
   Month   Day   Year

3. Permits for elevators and other building equipment as required by local authorities have been obtained. Copies of such permits were provided to Maintenance for display and to the Business Office for inclusion in contract records on _____
   Month   Day   Year

4. A copy of the General Building Permit was provided to the Business Office for inclusion in contract records on _____
   Month   Day   Year

5. A Certificate of Occupancy has been obtained from local building authorities as required and a copy was provided to the Business Office for inclusion in contract records on _____
   Month   Day   Year

6. The contractor has maintained accurate data relative to "as-built" construction. Two complete set of drawings marked in red ink showing all details of construction as actually performed which differ from details shown on the Contract drawings has been provided by the Contractor. Such "as-built" data has been approved by the undersigned or his designated representative and was furnished to Engineering on _____
   Month   Day   Year
   The other set was provided to Maintenance on _____
   Month   Day   Year

7. Waivers of Right of Liens have been submitted by the Contractor in accordance with applicable contract provisions and were furnished to the Business Office for inclusion in contract records _____
   Month   Day   Year

8. Final acceptance of the work performed under this contract was accomplished on _____
   Month   Day   Year
   by _____
   Name

_____   _____
Signature   Date

If Items 2, 3, or 5 do not apply to this contract, enter the words "not applicable" on the Month, Day and Year line.

FIGURE 11-1

SAMPLE COMPLETION NOTICE

FIGURE 11-2

SAMPLE EVALUATION FORM

CONTRACTOR PERFORMANCE
EVALUATION REPORT

JOB NO. _____

CONTRACT NO. _____

NAME OF CONTRACTOR

TYPE OF CONTRACTOR

DESCRIPTION AND LOCATION OF WORK

CONTRACT DATA

TYPE OF CONTRACT

AMOUNT OF BASIC CONTRACT
$ _____

DATE OF AWARD _____

TOTAL AMOUNT OF CHANGE ORDERS
NUMBER _____ TOTAL $ _____

ORIGINAL CONTRACT
COMPLETION DATE

NET AMOUNT PAID CONTRACTOR
$ _____

REVISED CONTRACT
COMPLETION DATE

☐ FORMAL BID

☐ SMALL CONTRACT

☐ SOLE SOURCE

☐ OTHER (SPECIFY) _____

DATE WORK ACCEPTED

PERFORMANCE EVALUATION OF CONTRACTOR

	Outstanding	Satisfactory	Unsatisfactory
Quality of Work			
Timely Performance			
Cooperation			
Effectiveness of Management			
Compliance with Safety Requirements			
Compliance with On-Site Requirements			
Participation in Cost Reduction			
Overall Evaluation			

PERFORMANCE EVALUATION OF SUBCONTRACTOR

List Subcontractors with Outstanding or Unsatisfactory Evaluations.
All Others Will Be Accepted As Being Satisfactory.

	Outstanding	Unsatisfactory

IF OUTSTANDING OR UNSATISFACTORY EXPLAIN EVALUATION ON REVERSE SIDE

EVALUATED BY _____ DATE _____

12 How to Cut Maintenance Costs Through Contracting Janitorial Services

If a maintenance manager contracts for janitorial services, he can achieve substantial cost savings by adopting the techniques described in this chapter. Janitorial contracting in large-scale facilities is a tough, competitive business. A maintenance manager who purchases janitorial services should be mindful of this fact and be prepared to protect his organization from the various competitive practices whereby janitorial contractors seek to maximize their profits at the buyer's expense. Basically, contracts for janitorial services consist of the following two types:

1. Cost-Per-Square-Foot Contracts
2. Purchased-Labor Contracts

COST-PER-SQUARE-FOOT CONTRACTS

Under a cost-per-square-foot contract, the contractor is paid a fixed rate for each square foot of facility for which he provides janitorial services. His price is predicated upon supplying the services as stipulated in the specifications. In some instances the buyer prepares his own cleaning specifications and conducts competitive bidding. In other instances, the customer negotiates a contract and uses the specifications provided by the contractor. A cost-per-square-foot contract is sometimes referred to as a fixed-price or fixed-rate contract because the rate per square foot is multiplied by the total square feet to arrive at a monthly price for the services.

BENEFITS

The major advantage to a cost-per-square-foot contract is its simplicity for the buyer. For a stated amount of money, a contractor agrees to do the work and furnishes all the labor, materials, and equipment necessary to perform the work. For the buyer, this type of contract requires the single process of letting one contract. He has eliminated the processes of buying supplies, hiring janitors, and acquiring cleaning equipment, which could necessitate a capital investment.

DRAWBACKS

The major disadvantage to a cost-per-square-foot contract is that the buyer lacks control over performance of the work. The contractor has every incentive to maximize his profits by cutting the size of the crew and using poor-quality, lower-cost materials and less than satisfactory equipment. Here are some examples of how cost-per-square-foot contractors have applied their ingenuity.

1. Crew-Shorting. The most common occurrence in cost-per-square-foot contracting is the practice of reducing labor costs, which is achieved by not replacing crew members who are out sick, on vacation, or who just don't show up for work. As a result, a smaller crew is always on the job than that which is really required to do the work, and all of the work is never done as often as it should be. Another crew-shorting technique is to reduce the number of leadmen in proportion to the number of janitors on the job, which lessens the amount of supervision given to the work and leads to poorer workmanship. In the airline terminal buildings of an airport, there was a contractor who used one leadman to cover the work crews for three different contracts when each buyer was supposed to have his own leadman, which constituted a 66 percent savings in leadman costs for the contractor. Each of the three customers, however, was paying for supervision that was not being provided.

2. Low-Quality Materials. In a large office complex, aisle areas were covered with resilient floor tile that was subject to heavy foot traffic. A refinishing schedule was established to periodically replace the finish and protect the tile surface. However, the finishing material began to wear off before the prescribed time for refinishing. It was discovered that the contractor had changed to a lower-quality finishing material that he had had specially manufactured at a cheaper price. These savings were not being passed on to the customer, and the surface life of his tile flooring was being severely shortened.

3. Poor Equipment. In a new county administration building, large areas were carpeted for a posh effect. Within nine months the carpeting was beginning to lose its elegance, and the wear could not be attributed to the quality of the carpeting or to heavy foot traffic. It was discovered that the vacuum cleaners used by the contractor had beating brushes mounted on plastic rather than metal rollers. The plastic was subject to scratching and denting. These knicks on the plastic roller were cutting the carpet fibers and substantially shortening the life of the carpet. The contractor was buying the soft-plastic replacement brushes at a substantially lower price than a higher-quality, hard-plastic or metal roller. These cost savings were not being passed on to the buyer, and the carpeting was being destroyed.

4. Using the Contractor's Specifications. The practice of accepting the contractor's specifications as part of the contract can also be a pitfall. The contractor has a strong incentive to work the specifications to his advantage. The specifications often contain several clauses for "extra work," which are not included in the fixed price. Substantial additional sums have to be paid to get all the necessary work accomplished. One airline experienced this problem in its food service area, which was excluded in the cleaning specifications and came into the "extra work" category. This practice raised total contract costs by 15 percent.

PURCHASED-LABOR CONTRACTS

Under a purchased-labor contract, the contractor is paid a fixed rate for the hours his personnel work. The work to be performed is prescribed in a set of specifications.

The hourly rate may be established through competitive bidding or by negotiation. Although this type of contract is termed purchased-labor, the contractor may also provide equipment or equipment and materials in addition to the labor. When equipment or materials are provided, their price is included in the hourly rate.

BENEFITS

The major advantage of a purchased-labor contract is that it eliminates the contractor's incentive to practice crew-shorting. He is paid only for hours during which his personnel are on the job site. Since a portion of the rate covers overhead and profit, the contractor actually loses money when his entire crew is not working. If the buyer furnishes the materials, the opportunity for the contractor to furnish low-quality supplies is also eliminated.

DRAWBACKS

One drawback to a purchased-labor contract is that the buyer has to establish some type of log and oversee the signing in and out of the contractor's personnel. Otherwise, the buyer has to rely upon the contractor's integrity with respect to timekeeping. When the buyer oversees the time log, it is normally done by a security officer who is located at an entrance duty station. Since the security officer is already there, the buyer's task of maintaining a time log is readily accomplished by this addition to the officer's assigned duties.

COST-PER-SQUARE-FOOT VERSUS PURCHASED-LABOR CONTRACTS

The method of contracting for janitorial services is basically a question of selecting the most effective means of controlling the costs and quality of janitorial services. Quality is measured in terms of how well the contractor adheres to the specifications. Cost is measured by how much the buyer has to pay the contractor for the work performed.

Contracting on a purchased-labor basis is advantageous in that it provides assurances that a minimum is paid only for those hours that contractor personnel actually log on the job site.

The purchased-labor approach is distinctly different from the cost-per-square-foot contracting method. To fully understand the principal differences between the two contracting methods, it is important to recognize the basic operating practices of a janitorial contractor under the rate-per-square-foot scheme. These are:

a. The contractor will contract for any rate that will assure him of winning the job and agree to the attendant specification levels.

b. Having been awarded the contract, the contractor will then use his management ingenuity to maximize his profits by (1) using understaffed crews and (2) omitting a certain percentage of square feet each day from the schedule.

Under a cost-per-square-foot method, the contractor is paid for work that has not been done. He is motivated not to perform to the specifications in order to increase his profits in the highly competitive market for janitorial services.

In comparison, the purchased-labor method eliminates any incentive for the contractor to practice crew-shorting. Further, there is no motivation for him to omit any service prescribed in the specifications, since he is paid only for those labor hours which are actually logged by his personnel.

The cost-effectiveness of the purchased-labor method can be demonstrated by the following examples:

1. One company used the fixed-price method for contracted janitorial services that cost $111,588 per year. A shift to purchased-labor contracts reduced these costs to $93,348. The net savings per year exceeded $18,000.

2. One maintenance department used the fixed-price method with a unit cost for janitorial service of $0.0550 per square foot per month. The department then changed to purchased-labor contracts. Nine years later, with contract janitorial labor rates inflated over 50 percent (from $1.83 to $2.82 per hour), the unit cost for *better* service, requiring less buyer administration, was $0.0527 per square foot per month. In addition, this $0.0527 rate included the extra requirements imposed by computer areas, machine shops, and manufacturing clean rooms that had been added to the facilities since the change to purchased-labor contracting. Excluding these extras, the rate for standard janitorial service was approximately $0.0487 per square foot per month.

This change in unit costs may seem insignificant until the rates are applied to the square footage of inventory involved, which was in excess of two million square feet. Using a 2,000,000 square feet figure, the total monthly costs using the fixed-price method equalled $110,000 (2,000,000 x .055). Under the purchased-labor method, even after nine years in an inflationary era, the total monthly cost equalled only $105,400 (2,000,000 x .0527). This figure represented an annual savings of $55,200 ($4600 x 12 months). Excluding the extra requirements, the total monthly cost nine years later equalled $97,400 (2,000,000 x .0487), which was an annual savings of $96,000 ($8000 x 12 months).

PROVIDING MATERIALS

When the buyer furnishes the materials for either a cost-per-square-foot or a purchased-labor contract, he incurs the administrative costs of procuring, storing, and issuing the materials. For large maintenance organizations, these tasks are usually added to the workload of the existing staff, which performs these functions for other maintenance supplies. Since a janitorial contractor normally includes these administrative costs in the overhead portion of his charges, the contractor's rates should be lower when he does not have to furnish the materials. The savings to the buyer derived from the lower rates should offset the increase in his administrative costs incurred by his providing the materials.

When the maintenance manager provides the materials, he also avoids the costs of the contractor's profit markup for furnishing the materials. This mark-up is generally in the range of 10 percent of the contractor's purchase price. If the maintenance manager is able to purchase the materials for the same prices paid by a janitorial contractor, he has the opportunity to make this savings in material costs. One maintenance manager who provided the materials was able to document a savings of over $15,000 a year by avoiding the contractor's markup.

When the maintenance manager furnishes the materials, he not only can control the quality of the materials used, but also can reap the benefits of increased efficiencies and better purchase prices. The following are some examples.

1. The maintenance department for a multibuilding site completely eliminated the warehousing and distribution functions on paper goods by establishing a schedule of incremental deliveries from the paper goods supplier directly to the janitorial stockrooms in each of the buildings. This technique yielded a savings of approximately $6000 per year.
2. A maintenance department for a large facilities complex in Southern California procured its cleaning supplies from a local chemical company rather than buying national brand names. The prices of the local chemical concern were substantially lower in all instances, because the advertising and distribution costs inherent in the prices of the national brands were eliminated. Using this method, the maintenance department acquired the materials at a savings in excess of $4500 per year.
3. The maintenance department for an airline terminal building switched to providing materials to its janitorial contractor. Contracts with a local chemical company provided a net savings of $30,000 per year.

CONTROLLING EQUIPMENT USED

It is necessary to have some control over cleaning equipment in order to assure that the most efficient equipment available is being used in the cleaning operations.

Buyer-Furnished Equipment

Maximum control of the equipment used is achieved when the buyer provides the equipment. Since he is furnishing it, he buys or leases what he thinks is best for the cleaning operations to be performed. However, there are four definite drawbacks in providing the equipment. These are:

1. The total capital investment required to own the equipment may be substantial.
2. Optimum utilization of the equipment may not be achievable in the buyer's facilities.
3. The cost of repairs is borne by the buyer, even though the repairs may be attributable to abuse by the contractor's personnel. If the contractor provides the equipment, he pays for any excessive repair costs due to employee negligence.

4. Change-over to more efficient equipment is slower, because the existing equipment is retained until its useful life has expired.

Because of these disadvantages, the normal practice is to have the contractor provide the equipment.

Contractor-Provided Equipment

The major advantages of having the contractor provide the equipment are the elimination of the four drawbacks involved in providing the equipment yourself. It is still possible to achieve control over the selection of equipment used and timely replacement of broken items through appropriate contract clauses. The following are three sample clauses to achieve control.

1. The contractor shall provide, at his expense, all power tools, machines and equipment necessary to perform the work as specified. Only industrial-type equipment is to be used.
2. The contractor shall furnish a recommended list of equipment to the buyer for review. All power tools, machines and equipment necessary to perform the specified work are to be approved by the buyer.
3. All equipment must be maintained in first-class working condition satisfactory to the buyer, and spare equipment must be available for replacement of broken items within twenty-four hours.

PREPARING CLEANING SPECIFICATIONS

There is a definite need for detailed cleaning specifications. The ability to control the quality of the janitorial program is provided by the specific detailing of the work to be done. This section presents some of the techniques which should be applied in developing specifications.

USING A BREAKDOWN BY TYPE OF SPACE

The purpose of the breakdown by type of space is to provide the janitorial contractors who will bid on the work with detailed data on the square footages of the various types of areas for which janitorial services will be required. This breakdown should be done in an effort to maximize the bidder's visibility and thereby reduce the amount of contingency which would be included in his quotation. Figure 12-1 is an example of a breakdown by type of space for an airport terminal building. In this example the major classification is by the type of flooring. Within each flooring classification there is a subdivision based upon the use of the area. Note that in the Vinyl Asbestos Floor Section some of the employee operations areas have been designated as "heavy use." This designation was made because rooms in this category will require more

TERMINAL BUILDING
JANITORIAL REQUIREMENTS
SUMMARY OF AREA BY TYPES

	SQ. FOOTAGE	SUBTOTALS
Exterior Concrete		
Landside Level I	40,360	
Landside Level II	41,100	
Airside Level I	62,769	144,229
Interior		
Concrete Flooring		
Transit Stations	8,354	
Baggage Check-In Areas	1,116	
Employee Operations Area	5,668	
Special Access/Storage Areas	39,445	
Unassigned Areas	18,272	
Restrooms—Employee	517	
Janitor's Closets	594	73,966
Vinyl Asbestos Flooring		
Employee Operations Areas	2,749	
Employee Operations Heavy Use Areas	6,063	
Special Access/Storage Areas	2,065	
Corridors and Vestibules	2,866	
Employee Restrooms	253	
Janitor's Closets	113	14,109
Carpeted Flooring		
Employee Operations Area	26,739	
Special Access/Storage Areas	1,704	
Corridors and Vestibules	2,247	
Computer Areas	1,869	
Employee Food Service	2,317	
Public Areas	134,671	
Elevator Lobbies	583	
Nursery	468	170,598
Ceramic Tile		
Restrooms	8,971	
Janitor's Closets	30	
Executive Passenger Club	9,215	18,216
Paver Tiled Public Areas	26,114	26,114
Quarry Tiled Employee Food Service Area	834	834
		448,066

FIGURE 12-1

SAMPLE BREAKDOWN BY TYPE OF SPACE

policing and more stringent floor care. Also, note the separate category for corridors and vestibules that will also require more floor care.

Figure 12-2 is a sample page of the specification format which uses the classification system shown in Figure 12-1. This format provides columns for describing the area, the applicable room number, and the square footage of the area. The Policing column is used to designate areas in which policing will be required.

USING FIVE-DAY NIGHTLY SERVICE

Figure 12-2 has a Daily Service column for the purpose of designating those areas in which only five-day nightly service is required. It should be recognized that a reduction in daily service in some areas from seven-day nightly service to five-day nightly service may not lead to any appreciable reduction in costs. A reduction in price or a savings will not be achieved unless the total area being reduced to five-day service

---

JANITORIAL REQUIREMENTS
EMPLOYEE OPERATIONS HEAVY USE AREAS—VINYL ASBESTOS FLOORING

Area Description	Room No.	Sq. Feet	Policing (1)	Daily Service (2)	Remarks
Coffee Room	2109	64	Yes		
Locker Room	2202	128		5	
Lunch Room	2203	263	Yes		
Locker Room	2348	120		5	
Lunch Room	2349	270	Yes		
Lunch Room	2417	388	Yes		
Locker Room	2418	72		5	
Lunch Room	2523	252	Yes		
Locker Room	2524	112		5	
		1,669			

Total Employee Operations Heavy
Use Areas—Vinyl Asbestos Flooring 6,063

(1) No policing required unless noted as Yes in this column.

(2) Service is required 7 days per week unless a 5 appears in this column, indicating 5-day service.

FIGURE 12-2

SAMPLE SPECIFICATION FORMAT

results in a reduction in headcount on the two nights for which no nightly housekeeping service is to be performed, since a janitorial contractor's employee is paid for eight hours whether or not there is sufficient work for the employee to perform during the eight-hour shift.

STATING FREQUENCIES IN JANITORIAL SPECIFICATIONS

In specifications for janitorial activities other than nightly housekeeping, all frequencies shall be expressed in terms of weeks, rather than months or quarters. This specification should be made to eliminate the slack time that results when activities are performed on a monthly basis, which averages to four and one-third weeks per month. Stipulating frequencies in terms of weeks provides a tighter and more explicit schedule that should be adhered to by the janitorial contractor.

STATING FREQUENCIES IN POLICING SPECIFICATIONS

Frequencies for policing requirements should be specifically expressed in terms of number of times per shift, instances of inclement weather, and so on. Such terms as "continuous" should not be used because they are too vague to be meaningful or

---

**JANITORIAL REQUIREMENTS
POLICING ACTIVITIES**

Public Areas and Employee Operations Heavy Use Areas—Police Twice Per Shift

1. Empty all waste baskets and other trash receptacles.
2. Empty and damp wipe ash trays and clean all sand urns.
3. Reposition furniture as required.
4. Pick up debris on floor. Remove gum and other foreign matter, and spot clean by damp mopping paver tile or composite flooring or spot cleaning carpeted areas.
5. Sweep paver tile or composite flooring in corridors, vestibules, lobbies, and other heavy foot traffic areas.
6. During inclement weather damp mop entrances to lobbies and lay strip carpeting provided by client.

Public and Employee Rest Rooms—Police Public Rest Rooms Twice Per Shift; Employee Rest Rooms Once Per Shift

1. Clean and refill all dispensers.
2. Empty all trash receptacles.
3. Sweep floors and spot mop.
4. Clean all mirrors, hand basins, and pullman counters.

FIGURE 12-3

SAMPLE FORMAT OF POLICING REQUIREMENTS

enforceable in dealing with a janitorial contractor's performance. Further, vague terms do not provide a bidder with data upon which to adequately formulate a headcount required per shift for policing activities.

DESCRIBING THE WORK TO BE DONE

An example of a format to describe the specific requirements for policing is shown in Figure 12-3. A sample format used to indicate the janitorial requirements is shown in Figure 12-4. Note that the statements are brief and start with action verbs for the work to be done.

EMPLOYEE OPERATIONS AREAS—COMPOSITE FLOORING

FUNCTIONS	FREQUENCY
1. Sweep all corridors, vestibules and all other flooring with a treated mop, or push broom covered with a treated cloth. Remove any spillage, gum, and other foreign matter and damp mop.	Each Night*
2. Empty and damp wipe ash trays.	Each Night*
3. Empty all waste baskets and other trash receptacles.	Each Night*
4. Clean and sanitize drinking fountains.	Each Night*
5. Dust and spot clean all furniture and fixtures such as desks, tables, chairs, file cabinets and ledges, partitions, baseboards, etc. to hand height. Spot clean any glass on desk tops.	Each Night*
6. Damp wipe all telephones.	Each Night*
7. Remove fingerprints and spot clean partition glass and doors.	Each Night*
8. Strip and refinish corridors, vestibules, and employee operations heavy use areas.	Once Per Week
9. Spot clean walls and paneling.	Once Per Week
10. Clean and sanitize telephones.	Once Per Week
11. Damp wipe all visual displays.	Once Per Week
12. Clean blackboards and dust chalk rails.	Once Per Week
13. Strip and refinish all flooring.	Once Every Four Weeks
14. High dust above hand height walls, air intakes and outlets, light fixtures, sills, ledges, locker tops, vending machines, etc.	Once Every Four Weeks
15. Wash interior and exterior of all trash receptacles and replace plastic liners.	Once Every Four Weeks
16. Clean all baseboards.	Once Every Four Weeks
17. Clean and polish door handles, push plates, and kick plates.	Once Every Four Weeks

*Seven days per week unless otherwise noted in the Room Listing for this type of area.

FIGURE 12-4

SAMPLE FORMAT OF JANITORIAL REQUIREMENTS

FLOOR FINISHING TECHNIQUES

In the past, use of a buffable wax has been the standard practice for most composition flooring (resilient or vinyl/asbestos tile). Here are some tips on eliminating this practice and saving labor dollars.

USING SPRAY BUFFING

Eliminating the practice of spray buffing composition floor has proven to be a cost-effective measure, particularly in heavy foot traffic areas. Spray buffing normally consists of the following sequential steps:

- Sweeping a floor
- Wet mopping a floor
- Spraying a light coat of wax on portions of the floor where the existing finish coat is wearing out
- Buffing the floor to obtain a luster finish

The practice of spray buffing has the following disadvantages:

- The spray coat of finish is applied to a floor surface which has not been thoroughly cleaned. Thus, the finish is intermixed with dirt which yields a suboptimal finish appearance.
- The spray coat contributes to a buildup of finish on the floor, which results in more time being required in the next stripping operation.

A more suitable practice, both from the standpoint of appearance and cost, is to strip and apply new finish to the floor on a scheduled basis. A nonbuffable, acrylic polymer floor finish that is extremely durable and black-heel-mark resistant will eliminate the need for spray buffing in heavy foot traffic areas such as corridors and aisles.

USING BUFF WAXING

Eliminating the practice of buff waxing composition floors has also proven to be cost effective. Buff waxing normally consists of the following sequential steps:

- Sweeping the floor
- Wet mopping the floor
- Buffing the existing finish to obtain a luster finish

This practice can be completely eliminated by applying a floor finish that retains its luster without periodic sweeping, mopping, and rebuffing. These floor finishes cost more money, but, by eliminating buff waxing activities, result in labor savings substantially greater than the increased cost for materials. Use of a nonbuffable, acrylic polymer floor finish that is extremely durable and black-heel-mark resistant will also eliminate buff waxing and the attendant labor costs.

APPLYING CARPET CLEANING METHODS

The three methods of carpet cleaning in most common use are dri-foam shampooing, steam cleaning, and wet shampooing. Each method has an appropriate use under the following conditions:

DRI-FOAM SHAMPOOING

Using dri-foam shampoo on carpeted concourses and other large carpeted areas has proven to be an effective cleaning method. Dri-foam carpet cleaning consists of the following sequential steps:

- Vacuuming the carpet
- Spot cleaning stains and removing foreign matter such as chewing gum.
- Applying a dri-foam
- Vacuuming the carpet

The practice of dri-foam cleaning has the following advantages:

- The carpeted area is virtually dry upon completion of the operation and is ready for use.
- The labor time required for the application and removal of dri-foam is substantially less than for wet shampooing. Wet shampooing normally requires *180 minutes per 1000 square feet compared to 45 minutes* for dri-foam cleaning of the same area.

WET SHAMPOOING AND STEAM CLEANING

The use of wet shampooing and vacuuming with a 17″ rotary machine, or using a small steam cleaner, is best in carpeted offices and other small carpeted areas.

Due to the size and maneuverability of a dri-foam machine, it is not suitable for use in small carpeted areas. Thus, carpeting in these small areas is normally cleaned using a 17″ rotary wet foam shampooing machine or a small steam cleaner. Either process takes about the same time.

Steam cleaning is promoted because of its alleged ability to penetrate further into the carpet and remove more dirt. It is probably more effective in cleaning deep pile or shag carpeting than a shampoo. The steaming process, however, may shorten carpet life by weakening the carpet fibers and backing.

When your carpeting has short piling, there is no need for the deep cleaning advantage of the steam process. The wet shampooing and vacuuming process is adequate for carpeted offices and other small carpeted areas which are not suitable for dri-foam cleaning. It does not create the risk of shortening carpet life.

USING STRIP CARPETS

Strip carpeting used in entry ways in inclement weather has definite advantages.

The use of strip carpets in rainy or snowy weather will protect entrance carpeting from water and dirt and will reduce carpet staining and added cleaning requirements.

Some buildings have paver tile in the entrance lobbies. The use of strip carpeting in inclement weather with paver tile is still advantageous in that the carpet strips will remove most of the water and snow from shoes and keep it from being tracked further into the building or onto the carpeted areas immediately adjacent to the paver-tiled lobby.

SEALING EXPOSED CONCRETE SURFACES

Exposed concrete areas should usually be sealed. Unless the concrete is sealed, cleaning is not efficient. A certain amount of dirt will always adhere to the surface no matter how often it is swept. More time is expended in sweeping the exposed surface which, over an extended period of time, adds more to janitorial labor costs than the cost of sealing the areas. Further, without sealing, the concrete will absorb oil and other liquids and leave stains which detract from the overall appearance of the areas. By sealing the exposed concrete in its baggage handling areas, one airline reduced janitorial labor costs by 10 percent, which resulted in an annual savings of approximately $7500.

13 Contracts for Groundskeeping and Landscaping Maintenance

As a start it is probably best to clarify terms. The term *groundskeeping maintenance* is used in a broad sense and refers to all maintenance work in all areas around buildings. It includes such items as sweeping of parking lots and roadways, maintenance of fencing and lighting, emptying trash receptacles, painting curbs and signs, road striping, and landscape maintenance.

Landscape maintenance is that part of groundskeeping which concerns work in planted areas, i.e., trees, shrubs, ground cover, and lawns. The work would include such items as fertilizing, trimming, pruning, watering, mowing, and maintenance of the sprinkler system.

Depending upon the nature of the organization, the level of emphasis on groundskeeping and landscaping will vary. For most colleges and universities, groundskeeping maintenance has always been given a high priority as part of projecting a suitable campus image. Today, even hard-line, profit-oriented manufacturing companies have changed the priority given to groundskeeping maintenance. Newer industrial parks are beginning to take on large amounts of landscaped areas around the buildings and planters in the parking lots in an effort to project an aesthetic image in the community, rather than have a facility that is a public eyesore. The days of a matchbox building surrounded by acres of asphalt parking area and a six-foot wire-mesh fence topped with barbed wire are disappearing. The surrounding community now generally expects any organization, government or private, to maintain facilities which lend some beauty to the area.

With this increased emphasis on the use of landscaping, the investment in trees, shrubs, ground cover plants and lawns can become substantial. Many industrial parks have well over a million dollars invested in trees and shrubs alone. Protecting this investment means substantial sums must be spent on landscape maintenance. It is no longer a simple matter of cutting a few lawn areas or trimming a few bushes when someone gets around to it. Because of this increased emphasis on landscaping, a large portion of this chapter is devoted to techniques in contracting landscape maintenance—how to reduce operating costs while protecting the investment and the quality of the landscaping program.

USING A FULL-SERVICE LANDSCAPE MAINTENANCE CONTRACT

Under a full-service contract the entire landscape maintenance operation is contracted. Normally, all materials, labor, and equipment are provided by the landscape contractor.

Full-service contracts are best used where the amount of landscape maintenance does not warrant a full-time in-house landscaping crew.

Figure 13-1 provides data on a comparative analysis of using a full-service contract versus having an in-house crew. The major disadvantage in contracting the program is that a contractor is essentially profit-oriented and the investment in landscaping may not be reliably protected over the long run. Short-term operating costs may be reduced through contracting, but long-term improvement and the initial investment in landscaping may suffer substantially. Lower cost may result in lower quality and a lack of protection of the investment in landscaping.

COMPARATIVE ANALYSIS SURVEY
OF
IN-HOUSE VERSUS CONTRACTED
LANDSCAPE MAINTENANCE

Using In-house Labor	Using Contractor
1. Usually a higher cost of operation.	1. Usually a lower cost of operation.
2. Usually a good quality of maintenance.	2. Usually a lower quality of maintenance.
3. Older techniques utilized and no high level of technical knowledge.	3. Usually a higher level of technical knowledge, with some application of newer, labor-saving techniques.
4. Oversold by material and tool suppliers and overly influenced by nurseries.	4. Minimum nutrients applied due to profit motive.
5. Dependable labor force with low turnover.	5. High turnover of inexperienced, unskilled personnel at worker level.
6. Landscape investment reliably protected.	6. No incentive for long term landscape improvement due to contract profit motive.

FIGURE 13-1

SURVEY OF LANDSCAPE MAINTENANCE

Full-service contracts are usually let on a competitive bid basis with the lowest bidder normally awarded the work. Unfortunately, the lowest bidder may be a contractor who hires low-paid, inexperienced workers, provides the least supervision, and uses a minimum of nutrients in order to maximize his profits. Also, he may not even employ a full-time, trained horticulturist, because of the added payroll cost.

There are techniques which may be employed as protection against a lack of quality. These are:

- A precise specification of the scope of work against which performance can be measured (Figure 13-2 is a sample outline for such a specification.)

- A thirty-day cancellation clause without the need to show cause and with no recourse to damages as a result of cancellation
- The requirement that bids be submitted on a unit cost basis
- The requirement that the bidder submit a capability or qualifications report including resumes of his technical and supervisory personnel
- A listing of current customers whose sites may be visited as being demonstrative of the quality of the contractor's work

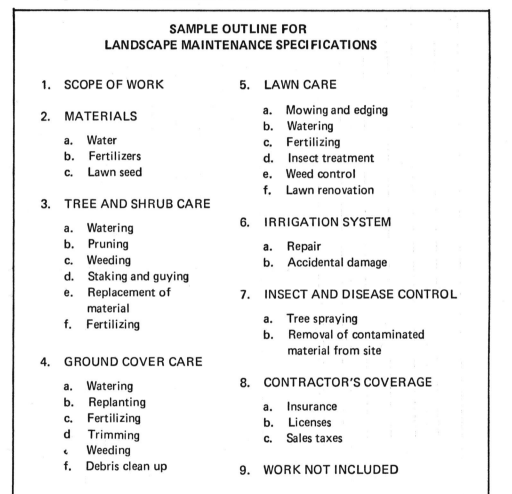

SAMPLE OUTLINE FOR
LANDSCAPE MAINTENANCE SPECIFICATIONS

1. SCOPE OF WORK

2. MATERIALS

   a. Water
   b. Fertilizers
   c. Lawn seed

3. TREE AND SHRUB CARE

   a. Watering
   b. Pruning
   c. Weeding
   d. Staking and guying
   e. Replacement of material
   f. Fertilizing

4. GROUND COVER CARE

   a. Watering
   b. Replanting
   c. Fertilizing
   d. Trimming
   e. Weeding
   f. Debris clean up

5. LAWN CARE

   a. Mowing and edging
   b. Watering
   c. Fertilizing
   d. Insect treatment
   e. Weed control
   f. Lawn renovation

6. IRRIGATION SYSTEM

   a. Repair
   b. Accidental damage

7. INSECT AND DISEASE CONTROL

   a. Tree spraying
   b. Removal of contaminated material from site

8. CONTRACTOR'S COVERAGE

   a. Insurance
   b. Licenses
   c. Sales taxes

9. WORK NOT INCLUDED

FIGURE 13-2

SAMPLE OUTLINE FOR SPECIFICATIONS

USING LANDSCAPE MANAGEMENT CONTRACTS

The purpose of a landscape management contract is to obtain the technical expertise of a trained horticulturist and the supervisory expertise of a trained field superintendent. The concept of this type of contract is somewhat new. In essence, it is an attempt to get the best of two worlds. An in-house maintenance crew is used in order to gain the advantages of a dependable labor force, control of quality and long-term protection of the investment in landscaping. At the same time, the technical expertise of a horticulturist and the most efficient management techniques are utilized to reduce labor costs. The major benefits are as follows:

- The cost of the contractor's profit on worker labor under a full-service contract is removed from the operation, since an in-house labor crew is used.
- Expert management is obtained for the landscaping program without having a full-time horticulturist or professional superintendent on the maintenance department staff. A landscape management contract is particularly useful when expert management is required, but the size of the program does not warrant the hiring of full-time professionals by the maintenance department.
- The expertise of a trained field supervisor is obtained in selecting the most effective and labor-saving equipment available for performance of various operations (e.g., mowing, verticutting, and spraying). Also, gardeners are trained on how to do quality work.
- Operations are planned and scheduled to achieve the optimum utilization of equipment and manpower.
- The trained horticulturist provides the infusion of the knowledge of agronomy into landscape maintenance programs. Environmental control agents such as selective herbicides, insecticides, and fungicides replace manual labor.
- Appropriate drainage, irrigation, and watering combined with proper fertilization, cutting, and pruning protect plant life and improve plant growth and appearance.

For one maintenance department landscape management contracting proved to be a highly effective method of reducing costs while protecting the investment in landscaping. Here were the results:

1. Through the use of labor-saving herbicides and improvement in equipment utilized, the headcount of the in-house labor crew was reduced from sixteen to nine landscape gardeners, which resulted in an annual savings of $70,000.
2. The in-house equipment inventory was updated to better match the scale of operation. Reel-type mowers were replaced with rotary, ride-type mowers which reduced mowing labor time by 20 percent. High-cost spraying equipment which had a low utilization rate was rented when required rather than purchased.
3. In addition to the reduction in costs, the use of the technical expertise of trained horticulturists led to an improved appearance of the site by better watering, drainage, and fertilization techniques.

There are many maintenance departments with large-scale landscaping operations which could benefit from the introduction of effective technical management and the adaptation of the labor-saving environmental control of agribusiness to ornamental horticulture. An effective method of obtaining these benefits is through the use of a landscape management contract. A major reduction in labor costs usually emanates from two sources. First, the use of herbicides combats weed and grass infestation and replaces manual labor. Second, the use of more efficient power equipment reduces operating labor time. At the same time, the use of insecticides and fungicides improves plant growth. As a result, there is a reduction in operating costs while the investment in landscaping is protected over the long run by a quality landscaping program.

CONTRACTING FOR OTHER GROUNDSKEEPING OPERATIONS

Contracting for groundskeeping services other than landscape maintenance is essentially like other buying. The three criteria of price, quality, and delivery should be applied in the contracting process. Price is best achieved through competitive bidding. Quality and delivery are achieved by specifications that are issued with the request for quotation and also made part of the subsequent contract. These specifications describe what is to be done, when it is to be done, and how often. After contract award, quality and delivery should be further insured by surveillance of the contractor to ascertain his adherence to the specifications.

As an example, let us discuss a contract for parking lot sweeping. Parking lot sweeping is probably the groundskeeping operation most frequently contracted on a routine basis. The high frequency is desired because most facilities do not have enough parking lot area to warrant the investment in a large power sweeper. A contractor can achieve a high utilization rate on his sweeping equipment, since he uses it for many customers, and he can offer the service at a lower total cost than if the work was done in-house.

A contract for parking lot sweeping is normally let on a competitive bid basis. The contract stipulates the time and frequency for the sweeping operation, for example, every Saturday and Wednesday between the hours of 5:00 a.m. and 6:00 a.m. To facilitate the sweeping operations, the selected time for performance of the work should be one when the parking lots are most apt to be empty of automobiles. To provide for contractor surveillance, the normal time selected is just prior to the beginning of the first shift. However, some sellers of sweeping services may not have equipment available at the desired time due to other commitments. A trade-off may have to be made to match the availability of equipment with the time best suited for the sweeping operation at the facilities. By stipulating the desired times for performance of the work in the request for quotation, the sellers are preconditioned to meet the ideal times for performance. Sellers who cannot meet the schedule will either not bother to submit a bid or will qualify the time for performance in their response. If the deviation from the desired time still permits performance of the work when the parking lots are most apt to be empty (e.g., thirty minutes after the end of the third shift in lieu of just prior to the beginning of the first shift), then the bidder's proposed schedule of performance may still be adequate. On the other hand, if the bidder's deviation from the desired schedule is not appropriate, the bid may be discarded as being unresponsive.

One maintenance department made the sad mistake of not stipulating a specific time for performance of the work. The first time the contractor's sweeper showed up was while employees were trying to park their cars at the beginning of the first shift. The congestion left something to be desired. The contractor took the hard-nosed position that this was the only time the equipment was available because of other customers who had contractual time requirements. Contract cancellation cost the company $350.

SECTION III–STAFFING

Food for thought:

A faithful employee is as refreshing as a cool day in the hot summertime. A lazy fellow is a pain to his employers—like smoke in their eyes or vinegar that sets the teeth on edge.

When you give a man a fish, you feed him for a day. If you teach a man to fish, you feed him for life.

POINTS TO CONSIDER

Staffing is the management task of manning, and keeping manned, the positions required in the organizational structure. It includes defining manpower requirements, recruiting, selecting and hiring candidates, compensating employees, and training employees to accomplish their tasks.

This section of the book contains three chapters on accomplishing staffing within a maintenance department. The first chapter deals with the staffing of craftsmen and trainees. The second covers supervisory positions. The last chapter is specialized and presents facets of the administration of formal apprenticeship programs for trainees.

There is a definite need for a well-conceived staffing program in a maintenance department. Without one, there will not be enough competent people around to execute your desired objectives in planning, organizing and controlling departmental activities.

14   *Staffing Craftsmen and Trainees for Results*

There has always been a requirement that a maintenance department hire and keep qualified craftsmen and trainees. Increasing labor costs necessitate a constant emphasis on recruiting, selection, and hiring.

Staffing, however, does not stop with the completion of the hiring process. Compensation, training and performance appraisal are also a part of the staffing function.

If compensation rates do not remain competitive, turnover is bound to increase. If there is no performance appraisal system, there is no formalized method whereby an employee can know how he is doing and where he is going. Turnover will increase again, because achievers will leave. Turnover is highly disruptive to operations. Further, for most organizations the average total costs of hiring an employee are in the vicinity of $1000. Therefore, turnover must be reckoned with on the basis of cost as well as inconvenience.

Training for maintenance personnel is required to improve the skills of the existing labor force. If equipment complexity increases faster than the technical capability of craftsmen, the gap will cause a decline in productivity. This gap can be kept closed by additional training of the work force to meet the maintenance requirements of the newer equipment.

This chapter presents some proven methods of how to determine the requirements for manpower and execute the processes of recruiting, selecting, compensating, training, and appraising employees.

DETERMINING MANPOWER REQUIREMENTS

The determination of requirements for manpower involves identifying how many people are required and what skills they should possess. As discussed in chapter 10, the planned use of outside contractors has a direct bearing on the manpower requirements of the internal work force, both with respect to the headcount and the craft skills required. Three basic levels of sophistication are generally found in determining manpower. These are:

- Informal
- Planned
- Formal

INFORMAL METHODS

In small maintenance departments involving ten craftsmen or less, informal methods are usually applied in determining manpower requirements. The scope of the operation does not necessitate or usually warrant a lot of effort in developing manpower requirements.

The maintenance manager assigns the work on a day-to-day basis. Unless there are unusual circumstances, such as a high level of absence due to sickness or injury, the small crew just chugs along doing their thing. When vacations occur, overtime or more use of relief men generally cover the situation. There is no need for a lot of planning effort. Determining manpower requirements is handled on an informal basis by the maintenance manager himself.

PLANNED MANPOWER SYSTEMS

In the middle level of sophistication in determining manpower requirements two basic methods are generally used. These are:

- The historical basis
- The adjusted historical basis

Either of these methods is usually found in a maintenance department with from twenty to fifty craftsmen. Some maintenance departments with 75 to 150 craftsmen also use the adjusted historical basis of determining manpower quite effectively.

Using the Historical Basis

The historical basis of determining manpower requirements essentially operates on the assumption that what was needed this year is the same as what will be needed in the future. The headcount and skills required for the internal work force are static.

The main advantage of this approach is its simplicity. The only planning required is for the replacement of retiring employees and a factor for labor turnover. Other than the need to fill vacancies resulting from retirements or terminations, the requirements remain in a steady state.

The main disadvantage to using this approach is a failure to gain any awareness of internal and external factors which may warrant changes in the requirements of the in-house labor crew. The situation may change, and without any knowledge of the change no planning will have been done to achieve the most cost-effective means of meeting the changed situation.

There are still a substantial number of large maintenance departments which operate on the historical basis. This lack of imagination on the part of their managers generally results in crisis planning when someone else recognizes that the world has changed and there are better ways to do the job of mixing the in-house labor crew and outside contracting. These managers are then placed in a defensive role of meeting the criticism and trying to justify what they have been doing.

It is seldom that a situation does not in fact change over time. Sole reliance on the historical method of determining manpower requirements over an extended period of years is not a good management practice.

Applying the Adjusted Historical Basis

This method provides for an awareness of change. Although past requirements are used as the base point in determining future needs in the internal work force, the method does provide for changing these requirements in the light of internal and external changes.

One maintenance department considers the examination of the following factors as essential in determining adjustments to historical data:

Internal changes in:

- Amount of square footage
- Types of square footage
- Quantity and types of plant equipment
- Age of facilities
- Plant operations—i.e., number of workdays or number of shifts supported

External changes in:

- Availability of contractors
- Changes in labor union philosophy
- Changes in technology

FORMAL SYSTEMS

Formal systems for determining manpower requirements are based on a planned level of effort. They involve fine-tuning manpower planning not only in the light of major internal and external changes, but also specific goals or objectives to be achieved in the coming year. For at least some craft skills, it usually involves the use of MICRO headcount planning, which was described in chapter 1.

Level-of-Effort Planning

Level-of-effort manpower planning requires the determination of the following items:

1. What is to be done and why
2. Where it is to be done
3. How it is to be done
4. When it is to be done
5. Who is going to do it

Each of the above items should be performed sequentially. When item 5 is reached (the who is going to do it), the decision to contract or perform in-house must be made. If the decision is to do the work in-house, the number of people and the craft skills required are derived from the first four items.

Because level-of-effort manpower planning is the most sophisticated, it is the most difficult and time-consuming method of determining manpower requirements. Generally, it is used only by maintenance departments with over 200 craftsmen, which are large enough to warrant having a planning staff. Automated management information systems are usually employed to facilitate the process.

There is no doubt that level-of-effort planning is a powerful method which permits the maintenance department to act rather than react. However, the method may not yield any better planning data than the adjusted historical method, while it takes substantially greater time and costs to reach the same conclusions.

Using Factor Analysis

The factor analysis method of determining manpower involves the use of a mathematical formula. The formula used will vary according to the mission of a given maintenance department or even a given craft within a department. The theory used is that the whole is equal to the sum of its parts. If a value can be determined for one of the parts, and this part can be expressed as a percentage of the whole, then the value of the whole can be determined. As a simple example of the use of factor analysis, let us use the planning of manpower requirements for maintenance machine tool mechanics. The following five symbols may be applied:

X = Total Men Required
B = Breakdown Maintenance
P = Preventive or Scheduled Maintenance
S = Scheduled Minor Repair
R = Rebuild or rehabilitation

The formula is:

$$X = B + P + S + R$$

Through MICRO headcount planning, let us assume that it has been determined that two maintenance machinists will be required for symbol P (preventive maintenance). (See chapter 1 for details on MICRO headcount planning.) Based upon history, it is also determined that P has equalled 30 percent of X. P (preventive maintenance) requires two men. These two men equal 30 percent of X, which is the total number of men required. The simple algebraic calculation to arrive at the total headcount required for maintenance machinists would be:

$$.30\ X = 2\ (men)$$
$$X = 2 \div .30$$
$$X = 6.66\ (men)$$

The 6.66 is then rounded to the nearest whole number which is seven. The total manpower requirement for maintenance machinists would be seven men. Since it is known that two men are required for scheduled maintenance, the remaining labor hours equivalent to five men are required to perform breakdown maintenance, scheduled minor repairs, and rebuild effort on machine tools.

Factor analysis is not a foolproof method for determining manpower requirements. It may prove to be a valuable planning tool if adequate historical data is available regarding the factors used. An automated information system greatly facilitates the use of factor analysis because of speed and accuracy in the manipulation of historical data.

USING JOB DESCRIPTIONS

As with the determination of manpower requirements, three levels of sophistication exist with respect to job descriptions for craftsmen. These are:

- Verbal
- Written
- Formal

VERBAL

Verbal descriptions are generally found in small, nonunion maintenance departments where all craftsmen are jacks-of-all-trades. The need to describe the job in writing has never occurred, because no organized labor union or personnel department has required it, and each man is expected to perform any task assigned.

Excluding the external pressures of a personnel department or a labor union, the use of verbal position descriptions can no longer be justified when a maintenance department begins to specialize by craft skill. It is now time to reduce descriptions to writing.

WRITTEN OPERATING INSTRUCTIONS

Written operating instructions tell a person what he is to do in accomplishing his job. Figure 14-1 is an example of a written operating instruction.

The main drawback in using operating instructions is that they imply, but do not directly indicate, the skills, training, and experience that are necessary to accomplish the tasks. While operating instructions may be adequate for use by existing employees, they are not sufficient for the purposes of recruitment and selection of new employees.

WRITTEN POSITION DESCRIPTIONS

Written position descriptions should indicate the skills, experience and training required to adequately perform a given job. They may be used in the recruitment and selection process, as well as in describing the responsibilities of a given position to existing employees. Figure 14-2 is an example of one type of position description. It describes the work processes that are to be performed, but fails to identify specifically the experience and training necessary to perform the work processes. Thus, it is not adequate.

OPERATION SHEET
STANDARD NO. 137
SCHEDULED AIR CONDITIONING MAINTENANCE

This is a second shift operation performed daily on weekdays (Monday through Friday) for all air conditioning equipment in the building designated on the Maintenance Work Order.

1. Inspect all air conditioning units and circulation pumps. Adjust as required for proper operation.
2. Record readings on log sheets.
3. Drain condensate in control and house air compressors.
4. Note any repairs made, any work not completed, or additional repairs required on the Maintenance Work Order.

Do not fill in the Date Complete block on the Maintenance Work Order unless all work has been completed when the Maintenance Work Order is returned to the foreman.

FIGURE 14-1
SAMPLE OPERATION SHEET

WORK PROCESSES
MAINTENANCE AIR CONDITIONING MECHANIC

A. Familiarity with shop practices, operation and use of equipment, hand and power tools, and materials.

B. Safety, work practice, education, and understanding of State Safety Orders.

C. Repair and function of equipment: absorption systems, pumps, heaters, coolers, compressors, condensers.

D. Air conditioning and refrigeration systems: install, check and calibrate, adjust and modify, usage of blueprints, sketches and manufacturer's specifications, and manuals to obtain locations and details.

E. Air conditioning and refrigeration equipment: maintenance, repair, trouble-shooting, overhaul; replaces, fits, and installs parts; diagnoses, checks and tests for leaks; charges system; sets and adjusts controls and regulators.

F. Instrumentation and control: piping, calibration and repair of instruments.

G. Chemical treatment, water treatment process, tests, water analysis.

H. Machinery operation: heat and cool to specified temperatures, maintain humidity, operating adjustments, thermostat work, reads gauges and instruments, records and maintains log, takes temperature and humidity readings.

I. Electrical controls, basic electrical theory relating to refrigeration systems.

J. Welding, silver brazing and soldering of piping.

K. Blueprint reading, heating and refrigeration codes, labor-material estimating, layout.

FIGURE 14-2
SAMPLE WORK PROCESSES

Figure 14-3 is an example of a position description that does provide experience and education requirements.

JOB DESCRIPTION

TITLE:

Maintenance Mechanic-A-Air Conditioning

PURPOSE/FUNCTION:

To perform the installation, maintenance, servicing, repairing, and rebuilding of air conditioning systems, refrigeration systems, and equipment.

PRINCIPLE DUTIES:

1. Determines by regular inspections the repair and maintenance work necessary to prevent breakdowns and major overhauls on air conditioning systems and refrigeration equipment. Recommends when equipment should be shut down for major overhauls. Plans sequence of operations and methods to use when making repairs, replacements, mechanical alterations, and overhauls, following trade practice and manufacturer's specifications.
2. Installs air conditioning and refrigeration systems. Checks and calibrates systems after installation and makes necessary changes, adjustments, and modifications to obtain desired results, using blueprints, sketches, and manufacturer's specification and manuals to obtain locations and details.
3. Maintains, repairs, troubleshoots, and overhauls complete air conditioning systems and refrigeration equipment; replaces, fits, and installs parts as necessary. Charges system with refrigerant, checks and tests for leaks, sets and adjusts controls and regulators, and checks operation after repairs.
4. Operates and adjusts air conditioning systems to heat and cool air to specified temperatures and to maintain proper humidity. Operates, defrosts, and services refrigeration systems and equipment. Makes operating adjustments as required. Checks and replaces thermostats. Maintains log or record of gauge readings and other operating data, and periodically takes temperature and humidity readings in given areas. Utilizes hand tools and air measuring devices such as thermometers, manometers, and draft gauges.

MINIMUM QUALIFICATIONS:

EDUCATION: One year of trade school courses in air conditioning and refrigeration theory and four year's experience, or five year's experience as an air conditioning mechanic.

EXPERIENCE—SKILL—KNOWLEDGE: Must be proficient in the use of mechanics' hand tools. Must be able to read and understand blueprints. Must be able to perform the above stated tasks in a satisfactory manner.

FIGURE 14-3

SAMPLE JOB DESCRIPTION

RECRUITING CANDIDATES

Policies of recruitment of personnel in the maintenance crafts are controlled to a marked degree by the local conditions existing for a specific maintenance department.

The recruiting process generally involves two types of individuals. One type is the trained craftsmen; the other is the trainee. The availability of trained craftsmen is largely governed by the conditions in the local labor market. In industrialized communities, the opportunities are obviously better. Some maintenance departments in newly industrialized areas have practiced importation to get an initial cadre of skilled journeymen. Thereafter, they have placed heavy emphasis on recruiting trainees in order to maintain the in-house maintenance crew. This approach has proven successful for them.

The challenge of obtaining qualified craftsmen or trainees can be difficult where a labor union contract provides for mandatory job posting. In unionized production plants, the maintenance departments usually have to select from candidates that become available through the bidding procedure, and normally must select on a seniority basis. The types of candidates made available can be improved only by trying to reach understandings with the union through improved union-management relationships.

If there are no mandatory recruiting practices dictated by a labor union contract, there are many other potential sources from which to get qualified applicants.

One source for obtaining qualified trainees that is frequently overlooked is within other parts of the organization. If the parent organization is a production plant involved in mechanical operations, competent recruits who are already partially trained can often be secured from the production departments. The other advantage of recruiting from within the organization is that the quality of the previous work performance of the candidate is more discernible than that of someone just walking in off the street. His employment record with the organization also provides some knowledge of his personal stability, teamwork potential, motivation, and ability to carry out assignments without constant supervision. Even if his current job assignment has not provided him with any substantial training in the craft for which he is being considered, at least you have a handle on his employment performance. A maintenance department in a paper manufacturing plant has recruited 40 percent of its millwright trainees from machine operators in the production departments. With few exceptions, these trainees developed into successful journeymen.

Acquaintances of employees are another source for both trained journeymen and trainees. This potential source is frequently not tapped, because no one bothers to tell employees that the job opening exists.

Trade schools are an excellent source for trainees. The graduating student has already demonstrated his educational preferences, as well as a certain amount of personal stability. Recruiting is usually conducted only from local schools. A maintenance department at a fairly remote location in Tennessee has been able to use trade schools in the nearest industrialized area as a good source for applicants.

SELECTING CANDIDATES

Four major tools may be used in selecting candidates. These are:

1. The application blank
2. The aptitude test
3. The interview
4. A physical examination

USING THE APPLICATION BLANK

The application blank should be used in all instances. From it you should be able to obtain the following information:

1. Name, address, age and sex
2. Previous work experience
3. Education and training background
4. References
5. An indication of interests
6. An indication of any physical disabilities

If you have this data on the application blank, it is a valuable tool in the selection process. With it you can screen out obviously unqualified candidates and avoid interviews. It is also the source document for conducting reference checks and verifying previous employment and work experience.

USING APTITUDE TESTS

The purpose of aptitude tests is to measure the potential for acquiring skills in an occupation. The tests are supposed to be given only to persons who have had no training or experience in the occupation under consideration.

Aptitude tests are frequently misused in two respects. First, the tests are administered to persons who have had training or experience, which is a misapplication of the test itself. The second misuse has to do with the test results. The test scores are often given too much emphasis in the selection process, rather than being considered in relation to all the other factors that may have a bearing on the individual's occupational success and satisfaction.

The use of aptitude tests is waning. This decline is largely attributable to the fact that the validity of the tests has become open to question, particularly with respect to minority applicants. Obviously, their use should be restricted to the selection of trainees. Candidates who are journeymen or craftsmen with previous work experience and training should not be given aptitude tests since the tests are not designed for them.

CONDUCTING SELECTION INTERVIEWS

The interview is the most time-tested of all the methods used in selecting candidates. There is no substitute for face-to-face contact in appraising the candidate and telling him what the job is all about. From the interview you should be able to garner information on personality, stability, education and training preferences, motivation, and interests. Additional information on previous education, training, and work experience may also be obtained in the interview.

USING A PHYSICAL EXAMINATION

The determination of physical capability or capacity should be handled by medically qualified people. A candidate may state that he likes manual labor or likes to work with his hands, but this is of little value if he has a chronic back ailment or a double hernia. The determination of physical capacity is normally the last step in the selection process. However, an indication of any incapacitating physical disability on the application blank should be a signal with respect to qualifying a candidate for continuation in the selection process.

APPLYING FACTORS IN SELECTION

The most commonly used factors in selecting candidates in the maintenance trades are:

- Motivation and interests
- Physical capacity
- Personal stability
- Aptitude
- Educational and training preferences
- Educational and training background
- Work experience
- Age

The extent to which these factors are considered in the selection process will vary markedly between the journeyman and the trainee recruit.

For the trainee candidate, motivation is frequently given the highest emphasis. The most preferable motivation is considered to be a genuine interest in the type of work, rather than a desire for security, prestige, or more money. The motivation factor should be explored in the interviewing process. Examination of the applicant's interests and education and training preferences is often used to gain insight into the applicant's motivation. Personal stability is also normally given heavy emphasis. Rather than measuring aptitude through a test, some maintenance managers prefer to indirectly evaluate it through high school or trade school achievement in shop classes which provide a background for learning the trade.

For journeyman candidates, age, education, and training preferences are not heavily weighted. The major emphasis is on previous work experience. The previous work experience is explored with respect to quality of performance, personal stability, and ability to carry out assignments without constant supervision, as well as the type of work done. This exploration is conducted by contacting previous employers, as well as through interviewing.

COMPENSATING CRAFTSMEN

When the labor force is unionized, the wages are set by contract and not adjustable by unilateral action. Flexibility is permitted in the compensation policy of the nonunion maintenance department.

Although wages are frequently not ranked as the most important factor by workers, the wage scale of a nonunion maintenance department should be maintained to stay equivalent to the going rates in the local area. An equitable wage scale increases the chances of holding down labor turnover.

Wage rates for government employees at city, county, and state levels and union scale rates are usually a matter of public record and readily attainable. Rates paid by private companies are less available, but can often be obtained from employment agencies or through information exchange programs between personnel departments of various firms. Periodic wage surveys should be made to assure that the pay scale of the maintenance department is competitive.

One nonunion maintenance department failed to conduct wage surveys. It discovered that it was no longer competitive when three of the nine journeymen electricians gave notice in the course of one week to take higher paying positions. This experience was a hard way to get smart.

TRAINING CRAFTSMEN

Training programs within maintenance departments range from full-scale state-approved apprenticeship programs to reliance entirely on exposure, supervisory coaching, and association with experienced craftsmen on the job. In between these two extremes there is a wide range of possibilities, including informal apprentice training, classes at equipment manufacturer's plants, classroom instruction at trade and technical schools, in-plant classes, and formalized on-the-job training.

The degree of formality of a training program is influenced by the same factors used to determine many other aspects of maintenance operations. These are:

- Availability of skills
- Size of operation
- Attitude of the labor group
- Value of results

A lack of availability of skilled craftsmen in the local labor marketplace increases the need and justification for a training program. Likewise, a change in methods or techniques may require additional skills. The only means of providing the necessary skills may be through a training program for in-house craftsmen.

The size of the operation influences the extent of formality of the training program. Only large maintenance departments can afford to initiate and maintain elaborate training facilities and programs. Smaller operations have to get by with substantially less.

The attitude of the labor group is important. If the in-house crew is unionized, a lack of cooperation on the part of the labor union could lead to a dismal failure. Even if there is no union, the support of the training program by craftsmen and supervisors is almost mandatory for success. Labor attitude should be seriously considered before any decision is made to initiate a training program.

The acid test for any training program is the value of its results. These results are demonstrated by improved maintenance operations either in staffing or performance. A training program should be a means to an end and not an end in itself.

Some maintenance departments that have instituted formalized training programs for existing journeymen have achieved startling results. Productivity increases have ranged as high as 25 percent. For one department, methods training at classes conducted at the equipment manufacturer's plant increased productivity by 20 percent for one of its crafts.

PERFORMANCE APPRAISAL

Over the years management has been bombarded by industrial psychologists, psychiatrists, and personnel consultants regarding performance appraisal. Some of the data is valid, and some of it is concocted by quacks who are trying to make a name for themselves. The quacks have led some maintenance managers to the point of throwing up their hands and discontinuing their formalized appraisal system. Other managers have never seen fit to install one.

There are five desirable reasons for having a performance appraisal system for craftsmen. These are:

1. To let the employee know where he stands.
2. To identify needs for individualized training or desired career development.
3. As a basis for rewards—wage increases, advancement in grade, service awards.
4. As a basis for discipline—discharge, demotion or static job status.
5. To develop a ranking.

INSTALLING A PERFORMANCE APPRAISAL SYSTEM

The installation of a performance appraisal system is best accomplished on a step-phase basis. The first step is to begin at the top by applying the system to your immediate subordinates. After they are thoroughly familiar with the objectives of the system, apply it to the next tier of supervision. Continue this process until you have arrived at the worker level.

At each level it is necessary to orient the group being evaluated to the method and objectives of the system at their level prior to conducting the actual appraisals. This orientation should prevent confusion and false anxieties.

The acid test for measuring processes is the degree to which the ratings jibe with real performance. If there is no match, the appraisal process is inadequate.

One maintenance department installed a new appraisal system without step-phasing. The system was implemented in one month for all levels of supervision and right down to the lowest paid worker. The industrial relations man who helped to install it got a promotion for being a hero, but the implementation was carried too far too fast. It took three years to get the system on-track. During that period, there was a lot of confusion and some frustration as members of the department gradually became oriented to what the system was trying to accomplish. The mess could have been avoided by a time-phased implementation plan that started at the top and worked down through the department levels.

CONDUCTING APPRAISAL INTERVIEWS

Successful performance appraisal programs provide for an interview between the appraiser and the appraisee. This interview should have the following objectives:

1. To provide feedback to the employee on where he stands, which includes covering positive as well as negative aspects of performance
2. To provide for a constructive discussion on where and how performance can be improved
3. To establish the needs for additional training for job enlargement or increased proficiency, for example, taking a course on motor control units so that the man can then begin to perform electrical maintenance on this type of equipment
4. To determine an employee's desires with respect to future opportunities for promotion or career development—i.e., supervisory progression, other job classifications, and what skills he will need to acquire to move up

Some foremen have expressed the opinion that an appraisal interview is unnecessary. They see the man daily and know how he is getting along. These foremen are mistaken. As one foreman so aptly put it: "There is a difference between seeing a man during the course of each day and having a performance interview. Normal workday discussions are about operations—what has to be done and when. In an appraisal interview you sit down and talk to the man about himself. If you don't take the time to do this, you never will really get to know the man or what his personal desires are, and he won't get to know what you think about how well he is doing his job and where he can improve his performance."

APPLYING CHECKPOINTS

There are some simple criteria that can be used in evaluating an existing appraisal system or in developing one to be used for craftsmen. The following six checkpoints have been successfully applied.

Use the Ratings

When the appraisal forms are completed, they should not be merely filed away. If they are not actually used for any of the five purposes of performance appraisal previously described in this section, the whole process is a waste of time.

Obviously, to let an employee know where he stands requires that the appraisal form be reviewed with him in an interview. Likewise, the form should also be reviewed with the employee with respect to its use as a basis for rewards or discipline.

Identifying the needs for individualized training for job enlargement or improvement is of little value unless there is follow-up to ensure that the employee actually gets the training.

A discussion of career development is rather useless unless there is a subsequent evaluation of desired opportunities and a path created. It should also be recognized that there are many journeymen who desire to remain just that. They have no wish to advance to supervisory or even leadman status. This being the case, their wish should be recognized. The employee who doesn't want to be promoted should not be prodded into taking a leadman or supervisory position. The job should go to a qualified man who actually wants the added responsibility.

To advance a journeyman who has no desire to take on a supervisory role can lead to unfortunate results. In one maintenance department, an experienced journeyman was forced into a leadman job. The man failed miserably. He was not comfortable in trying to execute his added responsibilities, and the crew was dissatisfied with the lack of direction they received from him. In the end he was demoted back to journeyman and he then left the company. Before his demotion, two men in his crew had already quit. The maintenance department lost three qualified journeymen as the result of trying to force a promotion on a man who did not want the job. When it was all over, the senior foreman who forced the promotion said: "One of the dumbest things I ever did around here was to make Bill take that job. He was a good worker and had been here so long that he knew each piece of equipment. He worked well with another man or as part of a turnaround maintenance crew. But he hated paper work and didn't like to give directions to a lot of men. He was a worker, not a leader. I should have recognized that."

Keep the Factors Work Oriented

An appraisal should concentrate on job performance, not personality. Job performance is a matter of behavior, or what a person does on the job. Personal characteristics or traits are those qualities that distinguish or identify a person (such as race), and may have little or no bearing on job performance. Characteristics such as emotional stability or attitude should not be included in the appraisal. Attitude is a matter of mental position, feeling, or emotion towards a job. Emotional stability is a matter for psychologists. A foreman is not a psychologist and should not be placed in the position of having to make a judgment on a craftsman's feelings or emotions.

In one instance, a journeyman took exception to a foreman's comment on his performance appraisal that the employee had a "lot of hang-ups." The journeyman won his case on the basis that the foreman was not qualified to make such a judgment.

UNILATERAL APPRENTICESHIP COMMITTEE
APPRENTICE EVALUATION

APPRENTICE NAME: _____

MAINTENANCE TRADE: _____

This form is to be completed on each apprentice at the prescribed time for each of his evaluations using the guide lines for rating each of the factors listed below.

FACTOR	GRADE LETTERS

	A	B	C	D	F
1. LEARNING PERFORMANCE					
2. QUALITY OF WORK					
3. ADAPTABILITY					
4. DEPENDABILITY					
5. WILLINGNESS TO LEARN					
6. CARE OF TOOLS AND EQUIPMENT					
7. ATTENDANCE AND PUNCTUALITY					
8. EFFECTIVENESS IN DEALING WITH OTHERS					
9. OBSERVANCE OF SAFETY PRACTICES					

GRADE LETTER RATINGS

A = Superior   B = Above Average   C = Average   D = Below Average   F = Unsatisfactory

ADVANCEMENT RECOMMENDED: Yes ☐   No ☐

Foreman's Recommendation and Comments: _____
_____
_____
_____
_____

Apprentice's Comments on Evaluation: _____
_____
_____
_____
_____

_____   _____
FOREMAN'S SIGNATURE   DATE

_____   _____
APPRENTICE'S SIGNATURE   DATE

SYSTEMS 4845

FIGURE 14-4
SAMPLE APPRAISAL FORM

Select Factors on the Basis of the Position

Some factors to be measured for a trainee or an apprentice are substantially different from those used in appraising a journeyman. Figure 14-4 is an example of an appraisal form used for an apprentice. Note that Item 1 (Learning Performance) and Item 5 (Willingness to Learn) are directly related to the trainee's acquiring knowledge. Item 6 (Care of Tools and Equipment) and Item 9 (Observance of Safety Practices) are also important factors in the development of an apprentice to journeyman status.

Now let's compare the nine factors listed in Figure 14-4 for an apprentice with the eight factors listed in Figure 14-5 for a journeyman. Note that Item 2 on the apprentice form is the same as the second factor on Figure 14-5 regarding Quality of Work. Attendance and Punctuality, which is Item 7 on the apprentice form, is Item 6 for the journeyman's factors. Item 8 (Effectiveness in Dealing with Others) on the apprentice form is endeavoring to appraise the same thing as Item 7 (Cooperativeness) for the journeyman. Item 3 (Adaptability) on the apprentice form is somewhat similar to Item 5 (Variety of Job Duties) for the journeyman. The other five factors on the apprentice form, however, are distinctly different from the factors used to appraise a journeyman.

Minimize the Number of Factors Rated

Too often the appraisal form becomes several pages long. While the number of

---

PERFORMANCE FACTORS FOR JOURNEYMAN

1. QUANTITY OF WORK—Ability to make efficient use of time and to work at high speed.

2. QUALITY OF WORK—Ability to do high grade work that meets quality standards.

3. ACCURACY OF WORK—Ability to avoid making mistakes.

4. KNOWLEDGE OF JOB—Understanding of the principles, equipment, materials, and methods that are directly or indirectly involved in the work.

5. VARIETY OF JOB DUTIES—Ability to perform different operations efficiently.

6. ATTENDANCE AND PUNCTUALITY—Extent of tardiness, lost time, and absences.

7. COOPERATIVENESS—Ability to get along with other workers and follow supervisory direction.

FIGURE 14-5

SAMPLE PERFORMANCE FACTORS FOR JOURNEYMEN

uses of the rating will exert an impact on the length of the form, it is questionable whether the number of factors appraised needs to consume more space than the front and back of one piece of paper. Any longer forms should be reviewed to assure that quantity isn't being substituted for quality. In one study it was found that thirteen factors could be reduced to seven with no significant loss of information. The seven factors are shown in Figure 14-5. Also note that the nature of these factors is essentially job-centered rather than personality oriented.

Quantify the Ratings

Quantification of the assigned rating for each factor appraised is a most effective means of facilitating a rating or ranking process. At the bottom of Figure 14-6 there are numerical values assigned to the grade letters for the nine factors of apprentice appraisal shown in Figure 14-4. Note that the first factor (Learning Performance) is weighted to equal 20 percent of the total evaluation. The other eight factors are equally weighted as 10 percent of the evaluation. A minimum passing grade to warrant individual advancement is 60 (i.e., at least a "C" rating on all factors). After an appraisal sheet has been completed on each apprentice, the total numerical score computed for each apprentice provides a rapid method for the compilation of a ranking list based upon merit. This same method of using numerical values can be applied to the ranking of craftsmen for merit advancement to another grade.

Many apprenticeship programs lay off on the basis of performance rather than seniority. In this instance, the ranking derived from the appraisal system is used to establish the layoff sequence based upon merit.

Maintenance departments with grade levels below the journeyman classification (such as a hierarchy of "C" workers, "B" workers, and then journeymen) also effectively use ranking when promotion from one grade to another is made on the basis of merit.

The appraisal system is rarely used to rank all of the journeyman in a craft. The usual reason for ranking journeymen is to determine a merit layoff sequence. Maintenance departments that are unionized, and a considerable number that are not unionized, however, lay off journeymen on the basis of seniority, not merit. Therefore, the appraisal system is rarely used for lay off ranking of journeymen.

Have the Ratings Reviewed

The rating review is normally implemented by the appraiser's immediate supervisor. This review process has several advantages. For example, when there are instances of serious disagreement, there is an evident need for further investigation of the appraisee's performance. Also, the review helps assure that the appraiser is properly applying the rating procedure.

Some advocates of performance appraisal believe that two appraisers should independently rate the same subordinate in order to provide some index of agreement that can be checked. The use of this system may be possible, but it is seldom

UNILATERAL APPRENTICESHIP COMMITTEE
APPRENTICE EVALUATION

GUIDELINES FOR EVALUATION BY FOREMAN

Apprentices are to be rated on the Evaluation Form for the nine factors indicated. Each factor is to be assigned one of the five letter grades from Superior to Unsatisfactory.   A = Superior   B = Above Average   C = Average   D = Below Average   F = Unsatisfactory

THE RATING ON EACH FACTOR IS TO COVER PERFORMANCE ONLY DURING THE PERIOD COVERED BY THE EVALUATION. The criteria for assigning a rating for each factor are as follows:

1. **LEARNING PERFORMANCE:** Consider demonstrated ability to learn the work processes covered by this apprenticeship.
   A = Learns very rapidly. Has shown rapid progress; has attained a high degree of proficiency.
   B = Learns more quickly than average: not quite satisfied with knowing minimum: does expected tasks with speed and accuracy.
   C = A competent worker. Has shown normal or average progress; has attained a reasonable or expected degree of proficiency.
   D = Has learned slowly; has not shown an expected degree of proficiency.
   F = A low ability worker. Has made very little or no progress; has shown little or no proficiency.

2. **QUALITY OF WORK:** Consider accuracy, neatness, and thoroughness of work in meeting quality standards.
   A = Consistently turns out work of a high degree of accuracy, neatness, and thoroughness. Quality is outstanding.
   B = Work is almost always accurate and neat. Checking seldom necessary. Usually finds and corrects occasional errors himself.
   C = Accuracy, neatness, and thoroughness of work generally satisfactory. Occasional checking required.
   D = Frequently makes errors. Frequent checking required. Work often lacks thoroughness and neatness, but usually meets minimum requirements.
   F = Continually makes errors. Complete checking of work required, work untidy, incomplete and frequently unacceptable. Below standards for job.

3. **ADAPTABILITY:** Consider how the apprentice adjusts to non-routine situations, consider resourcefulness, versatility, planning ability, and exercise of judgment.
   A = Highly resourceful and versatile. Unusual ability in dealing with new situations. Frequently offers good suggestions. Requires only a minimum of direction.
   B = Readily handles non-routine situations with resourcefulness and good judgment. Occasionally has new ideas. Occasional direction required.
   C = Usually exercises good judgment in dealing with various situations arising in job. Normal supervision required.
   D = Has difficulty in planning work and in handling non-routine situations. Needs frequent direction.
   F = Seems unable to adjust satisfactorily to variations from routine. Uses poor judgment. Needs constant supervision.

4. **DEPENDABILITY:** Consider acceptance of responsibility and application to job for present period of apprenticeship.
   A = Shows unusual sense of responsibility. Carries out instructions conscientiously and promptly. No follow-up required. Justifies complete confidence.
   B = Can almost always be relied on to carry out instructions and to complete assignments promptly. Follow-up seldom required.
   C = Generally follows instructions. Usually prompt in completing assignments. Occasional follow-up required. Usually can be relied on to fulfill job requirements.
   D = Often fails to carry out instructions; frequently loafs on the job. Needs occasional prodding to get work done.
   F = Has little sense of responsibility to job. Kills time. Invariably fails to follow instructions. Unreliable.

5. **WILLINGNESS TO LEARN:** Consider willingness to accept direction and supervision.
   A = Very congenial and cooperative. Very interested in work. Actively cooperates with supervision and in receiving instruction in work processes.
   B = Is cooperative. Shows above average interest in work. Readily accepts supervision and instruction and profits by it.
   C = Reasonably cooperative. Shows interest in work. Accepts supervision and instructions on work processes.
   D = Sometimes difficult to work with. Tends to be balky and to irritate others. Often indifferent. Tends to disregard instructions.
   F = Frequently uncooperative, touchy, unpleasant, or antagonistic. Shows little or no interest in work. Resists supervision or instruction in work processes.

6. **CARE OF TOOLS AND EQUIPMENT:** Consider care during present period of apprenticeship.
   A = Unusually outstanding in caring for tools and equipment.
   B = Can almost always be relied upon to properly care for tools and equipment.
   C = Usually careful in the treatment of tools and equipment.
   D = Often neglectful of tools and adherence.
   F = Unusually careless in the care and treatment of tools.

7. **ATTENDANCE AND PUNCTUALITY:** Consider performance during present period of apprenticeship.
   A = Almost never absent or late.
   B = Infrequently absent or late.
   C = Occasionally absent or tardy. Does not effect work.
   D = Frequently absent or late.
   F = Excessively absent or tardy. Affects work performance.

8. **EFFECTIVENESS IN DEALING WITH OTHERS:** Consider effectiveness during present period of apprenticeship.
   A = Exceptionally courteous and well mannered.
   B = Tactful and obliging; good self-control.
   C = Usually maintains courteous, effective relations.
   D = Lacks certain requirements of common courtesy; complaints occasionally received.
   F = Surly, touchy or quarrelsome; antagonizes requester's and fellow workers.

9. **OBSERVANCE OF SAFETY PRACTICES:** Consider adherence during period of apprenticeship.
   A = Unusually outstanding in following safety practices.
   B = Can almost always be relied upon to follow proper safety procedures.
   C = Usually adheres to safety practices.
   D = Often neglectful of proper safety practices.
   F = Unusually careless in observing safety practices.

NUMERIC VALUES OF GRADE LETTERS

Grade Letters for each of the nine factors evaluated equate to numeric values as indicated below. Note that the first factor (Learning Performance) is weighted to equal 20% of the total evaluation. The other eight factors are equally weighted as 10% of the evaluation. A minimum passing grade to warrant advancement is 60 (i.e., at least a "C" rating on all factors).

	A	B	C	D	F	
1. LEARNING PERFORMANCE	20	16	12	8	4	
2. QUALITY OF WORK	10	8	6	4	2	
3. ADAPTABILITY	10	8	6	4	2	
4. DEPENDABILITY	10	8	6	4	2	N U M E R I C
5. WILLINGNESS TO LEARN	10	8	6	4	2	
6. CARE OF TOOLS AND EQUIPMENT	10	8	6	4	2	V A L U E S
7. ATTENDANCE AND PUNCTUALITY	10	8	6	4	2	
8. EFFECTIVENESS IN DEALING WITH OTHERS	10	8	6	4	2	
9. OBSERVANCE OF SAFETY PRACTICES	10	8	6	4	2	
TOTALS:	100	80	60	40	20	

Header above table columns: GRADE LETTERS

FIGURE 14-6

SAMPLE NUMERIC VALUE

effectively practicable in the administration of a maintenance department. It usually takes concentrated management effort and planning to get a single appraisal done properly and on time, let alone two independent ratings.

USING COMPARATIVE JUDGMENT

The comparative judgment technique of performance appraisal consists of two elements. First, all craftsmen in the group are rated at the same time on each factor used in the appraisal. Second, each craftsman is rated in comparison with the other craftsmen in the group, rather than with some hypothetical set of standards.

To properly apply the rating of the whole group on one factor at a time, all journeymen in a given craft must be rated at the same time, which is distinctly different from having individual appraisals at different times, such as upon each employee's anniversary date. For this reason, some maintenance managers are opposed to trying to use the comparative judgment technique. Rating a crew all at once can be a time-consuming procedure compared to spreading the process out over a year.

Those in favor of the practice feel that it is worth the effort to do the whole job at one time, since the ratings will be more consistent and equitable. One manager puts it this way: "The only way to get comparative judgments is to cover all the men at the same time. Spreading the appraisal process out leads to inconsistencies. A foreman may be extremely busy one month and do only a cursory job of rating those people who are due for an appraisal during that period. When we plan a specific time period for the entire process, each of my foremen knows the job is to be done then and he had better be prepared to devote the proper amount of time to it."

WATCHING FOR PERSONAL BIAS

Each of us has some amount of personal bias. If there is to be equity in the performance appraisal process, one must be aware of this bias and guard against it. The following are some of the usual difficulties.

Halo Effect

This shortcoming consists of making a judgment of the whole on the basis of some part. Most workers are better at one ability than another. For example, a worker may be very accurate, but very slow. If you rate the worker low on all abilities just because he is slow, then you are applying the halo effect.

Leniency Error

This term describes the supervisor who rates everyone too high. The rater does not want any unpleasantness stemming from a face-to-face discussion with the worker about where job performance should be improved.

Central Tendency Error

This factor consists of the practice of considering everyone about average and avoiding necessary extreme appraisals. The rater who uses central tendency not only avoids any unpleasantness in discussing areas that may need improvement, he also fails to give recognition to achievers.

Incident Rating

This error is rating a craftsman on the basis of one good day or one bad day, or some single incident, rather than on the basis of the work done over the entire rating period. For example, an apprentice being considered for advancement to his third step received a low rating on safety practices from his foreman. When asked why the rating was given, the foreman replied that "the kid dropped his hammer off the ladder the first day on the job and darn near brained me." This one incident, which did not even occur in the period for which the rating was applicable, had formed the basis for a consistently low rating on this factor.

Stereotypes

This problem is the tendency to apply an inaccurate general guiding principle, such as appearance, race, or age, that is not relevant to job performance. In one instance a foreman gave a craftsman low ratings on all job performance factors. This rating did not jibe with the reviewer's impression of the employee. A discussion of the matter revealed that the foreman was really handing out a low rating on the basis that "the guy ought to shave off that beard."

15 *Staffing Supervisors for Top Performance*

There is a direct correlation between departmental performance and the quality of its supervisors. This chapter covers staffing of supervisors. As with staffing of craftsmen, staffing of supervisors entails more than the processes of recruiting, selection, and hiring. Compensation, training, and performance appraisal are equally important. Proven methods for accomplishing the staffing of supervisors are contained in this chapter.

DESCRIBING THE POSITION

There is an axiom that states that you cannot tell a man what is expected of him if you do not know what you want him to do. There is a definite need to thoroughly analyze each supervisor's job. After the analysis, the description of the job should be reduced to writing. This task may take some time, but it is worth it. Retention of facts by memory is faulty, and items of responsibility can get lost in verbalizing. People hear what they want to hear. The written word is less subject to interpretation.

In one experiment, four general foremen were read the job description for a supervisory position with which they were familiar. Each one was then asked to describe the job. None of them remembered all of the requirements of the job. Two of them creatively added responsibilities which were not even in the description! No wonder candidates who are interviewed by several persons for a job get different interpretations of what the job is all about. After they are hired, it is easy to see why they get conflicting descriptions of what they should be doing.

The need for a position description cannot be overemphasized. As discussed further in this chapter, the position description can be an effective tool in accomplishing the recruiting, selection, and compensation processes.

An adequate supervisory position description should contain three items. These are:

- A description of the job in relationship to the organization
- The principal responsibilities of the job
- The minimum education and work experience qualifications for the position

Figure 15-1 is an example of a position description for a supervisor of maintenance stock. Note that there are three separate sections to the position description covering: organizational relationships, principal duties and responsibilities, and qualifications. The duties and responsibilities section is a listing of nine specific duties or responsibilities. The tenth item is somewhat less specific and serves to indicate that additional duties may be assigned at the discretion of the immediate superior.

182

POSITION DESCRIPTION
SUPERVISOR OF MAINTENANCE STOCK

ORGANIZATIONAL RELATIONSHIP

Reports to the general foreman of administration. Coordinates with line foremen in the determination of material requirements. Coordinates with senior foremen in the determination of tool crib item requirements.

PRINCIPAL DUTIES AND RESPONSIBILITIES

1. Supervises personnel assigned to the material control, stockroom, and tool crib sections.
2. Responsible for procurement and inventory control, including ordering, receiving, issuing, and maintaining inventory records of maintenance material and the preparation of financial reports.
3. Responsible for tool crib operations including procurement, inventory control, and maintenance of tool crib items.
4. Conducts usage analysis and coordinates with line foremen in determining items classified as standard stock inventory, including the determination of minimum and maximum stock balances.
5. Coordinates with senior foremen in determining tool crib requirements, including specific tool crib items, quantities, and locations for issuance.
6. Processes purchase requisitions for procurement of capital equipment.
7. Assists senior foremen in preparing their annual expenditure plans for tool crib items and compiles the total requirements for inclusion in the annual budget.
8. Assists the business manager in preparation of the annual budget for maintenance materials.
9. Prepares and administers the budgets for the material control, stockroom and tool crib sections.
10. Conducts material analysis studies, prepares special reports, and performs other duties assigned by the general foreman of administration.

QUALIFICATIONS

Education

Associate in Arts Degree, Certificate of Achievement in Business Administration, or satisfactory completion of at least thirty hours of college level courses in subjects such as material control, purchasing, managerial and cost accounting, business management, and supervision.

Experience

Minimum of five year's experience in material procurement and inventory control operations, with at least two year's supervisory experience.

FIGURE 15-1

SAMPLE SUPERVISORY POSITION DESCRIPTION

RECRUITING SUPERVISORS

Three sources are generally used in recruiting supervisory candidates. These are:

- The maintenance department
- The parent organization
- The job market

PROMOTING FROM WITHIN

Probably as much as 70 percent of the supervisory personnel within maintenance departments are obtained by promotion of existing employees. This practice of promotion from within the department has some advantages. First, the employee is already familiar with departmental policies and operations. He merely is changing his role within a familiar environment. There should be less learning time needed before he can become independently effective. Second, employees in the department who desire to get into supervision see an opportunity for advancement. The practice of promotion from within is generally followed extensively, and only if there are no suitable candidates available are other sources utilized.

GETTING PEOPLE FROM OTHER DEPARTMENTS

Recruiting from within the organization has certain advantages. Even though the candidate may not be thoroughly familiar with the operations of the maintenance department, he is at least familiar with the policies of the parent organization.

A policy of cross-promotion of personnel between departments also has the advantage of serving to improve interdepartmental relationships through familiarity with each other's needs.

Another advantage is that the track record of a candidate from within the parent organization is more readily verifiable than that of the candidate from the open market. The availability of information on a current employee facilitates the selection process.

Recruitment from within the organization is most frequently followed within manufacturing plants, where production supervisors move over to the maintenance department. Since they know the machines and the maintenance support requirements, they can be effective immediately in their new job.

RECRUITING FROM THE OPEN MARKET

There are two reasons usually given by a maintenance manager as to why he recruits from the open market. The first is that there is no talent available within his own department or other departments of the parent organization. The second reason is that he wants new blood. Generally, there is a relationship between these two stated reasons. In essence, when a maintenance manager says he wants new blood, he means that he wants to shake things up, change some policies, and make things run differently. He wants new talent that will help him in this redirection of the department's activities. He feels that there are no suitable employees within the department or the parent organization who can do this, so he is going outside to get new team members.

Advertising in Newspapers

One of the most commonly used methods of recruiting outside the firm is newspaper advertising. Using the minimum qualifications stated in the position description as part of the advertisement helps to curtail the number of responses from unqualified people.

Newspaper ads may be open or blind. A blind ad has the respondents send their resumes to a post office box and does not identify the organization seeking personnel. The blind ad permits the screening of the responses before any time is wasted in personal contacts with obviously unqualified candidates.

Since the open ad identifies the firm, it usually results in phone calls and personal visits in addition to the requested mailed resumes. Since these visits and phone calls are normally handled by the personnel department, the maintenance manager is not wasting his time talking to unqualified people. The personnel department, however, is spending its time being courteous to people who are often unqualified candidates and who have failed to follow the instructions in the advertisement to mail in a resume. For this reason many personnel departments prefer to use blind ads.

Blind ads can lead to some surprises. One maintenance manager recruited for a supervisor using a blind newspaper advertisement. In the responses, he got a resume from one of his own foremen. The man was considered to be an outstanding performer and the manager was not even aware that the foreman was so dissatisfied that he was seeking employment elsewhere. The ad brought the situation to the surface where it could be worked out. (Incidentally, this particular maintenance manager did not believe in having a formalized periodic performance appraisal system, which may account for the fact that the foreman's dissatisfaction came as a complete surprise.)

Using Personnel Agencies

The use of personnel placement agencies has one definite advantage. Reputable agencies will have done some screening of their candidates, and applicants who are referred will usually meet at least most of the job qualifications. Giving the agency a copy of a written position description facilitates this process.

It should be recognized, however, that placement agencies make their living from the fees they collect either from the hiring firm or from the candidate. Quite naturally, they are prone to stress the attributes of their referrals and gloss over any shortcomings, because they do not get a dime until the hiring is accomplished. For this reason, many maintenance managers do not like to use job placement agencies and rely on newspaper ads as their recruiting method.

USING ASSOCIATIONS

A third potential recruiting source for supervisors in the open job market is national associations. The contacts are usually with the local chapters which, in turn, can work with the national headquarters or other chapters in providing potential candidates. Being a member of the association often facilitates the process. One of the primary reasons some maintenance managers retain membership in these associations is to seek out supervisory talent.

SELECTING SUPERVISORS

Some maintenance managers consider the selection of supervisors as one of the most important operations in performing the job of staffing. The effect of the selection is threefold. First, there is the effect on the person selected. Second, there are the relations with employees who will be under the new supervisor. Third, there are the employees of the maintenance department who were also candidates but did not get the job.

The supervisor's job is related to the organization structure. An organization functions as a group of coordinated sections. A supervisor is responsible for seeing that one of these sections performs its assigned operations. A supervisor acts with the authority of the maintenance manager on certain delegated functions. Supervisors should be selected and retained for their ability to apply the maintenance department's policies to each situation and arrive at substantially the same decision that the manager himself might have made.

The above may sound somewhat theoretical, but it really gets at the heart of who you should select. Supervisors who are successful apply the maintenance department's policies to situations and arrive at substantially the same decisions as those that would have been reached by the maintenance manager himself. Supervisors who cannot make the right decisions ultimately fail in the job. The failure is not always completely the fault of the supervisor. He must know what the policies are if he is expected to apply them. This matter is discussed further in the section of this book on the management function of directing.

In determining which qualified candidate to select, a maintenance manager for an operation involving over 200 craftsmen has found that he selects the man whom he feels is best able to control methods, pace, and quality of performance. These three items are what he deems the most important functions of a supervisor.

Another maintenance manager states his position this way: "A candidate may be the nicest guy you have ever known, been in the department a long time, and be a highly proficient craftsman. But if the guy does not have the minimum job qualifications, cannot handle paper work, or plan, staff, organize, and control the work of others, then he has no business being made a supervisor. Actually, to promote him is an injustice, because he won't hack it. Then you either have to demote him, fire him, or limp along while you support his shortcomings. The worst mistake is to limp along. If you crutch him, you are doing an injustice to the rest of your supervisors, because the time you spend crutching an unsatisfactory supervisor should be spent in directing your other members of supervision. You are making yourself unavailable to the rest of your supervisors, because you are spending time helping the unsatisfactory supervisor do his job. You may feel good about carrying the guy, but the injustice that you are doing to your other supervisors will ultimately have an adverse effect upon the operations of your department. When I select a guy to be a supervisor, I try to pick a man I feel will make decisions and see to it that his section performs its assigned responsibilities without having to be wet-nursed. If I make a poor selection, and I have made some, then I face up to it and refuse to carry a nonachiever. There is a distinct difference between directing a guy and doing his job for him."

COMPENSATING SUPERVISORS

There are two factors that should be considered in determining the compensation of supervisors. First, the salaries should be competitive with those paid by other organizations in the local area. Second, the salaries should be equitable in relation to the total wages paid to journeyman reporting to the supervisor.

DETERMINING COMPETITIVE POSITION

For the first factor, data is available from local government agencies on their supervisory salary scales. Personnel placement agencies are also a source of data. The personnel department of the parent organization may also have a wage survey interchange program with other organizations in the area. Here again position descriptions are a valuable tool that can be used in determining if the salary data is for equivalent positions. Job titles will vary, but with definitive position descriptions one can be assured that the jobs are comparable.

ESTABLISHING EQUITY WITH JOURNEYMAN WAGES

Data concerning equity in the compensation of journeymen is more readily available. The rates paid the journeyman in the department are known quantities. To arrive at the total wages paid, it is necessary to consider both the average straight-time and the premium-time earnings paid to a journeyman. It is usually the premium-time earnings that create inequities. The journeyman ends up making more than his supervisor because of the extensive number of overtime hours that he works at a premium rate.

The key to establishing an equitable total compensation plan for supervisors is to use the same base point as the pay plan for journeymen. This base point is hours worked. It is when the supervisor is not compensated for additional hours worked that the inequity arises. Some provision for supplementing the supervisor's salary for additional hours worked should be made. This provision is normally achieved through an extended workweek compensation plan that provides supplementary compensation for the additional hours actually worked by the supervisor.

Extended workweek compensation plans for maintenance departments vary. Some pay the additional amount at the base rate of the supervisor. Others add a premium to the base rate. Most plans set a maximum amount that can be paid during a given period. There is no pay for hours worked beyond the maximum. The reason for setting a maximum is to assure that the plan is not abused. An extended workweek pay plan should not encourage the practice of having an inadequate number of supervisors within the department.

TRAINING SUPERVISORS

Supervisory training should have two distinct objectives. These are:

1. Improvement of performance on the present job
2. Development for a future position

Training, as opposed to directing, involves some sort of formal classroom instruction. This training can be in-house or outside the organization.

TRAINING OUTSIDE

Training outside the organization normally is of two types. These are:

1. Formal classes of instruction at trade schools and colleges
2. Conferences and colloquiums

Formal classes of instruction usually involve an extended number of sessions and are taken by the supervisor on his own time. The subject matter may be technical or oriented to business administration. These types of courses can be beneficial both to improve performance on the present job or as development for a future assignment. However, it should be recognized that most courses offered are general in nature and only a part of the course content can be directly related to a specific organization. The amount of this type of training or education that is available is largely a matter of the geographical area. Maintenance departments operating in heavily industrialized areas will usually have a wide range of technical schools and colleges available. Those that do not must rely more on conferences and colloquiums as the major part of their outside training program.

Conferences and colloquiums are of a short duration, lasting from one to five consecutive days. They may be sponsored by companies, professional organizations, or training schools. The supervisor is normally paid for attending these sessions, since they are given largely during normal working hours. Conferences may also require travel and overnight lodgings, which are also reimbursable costs. Thus, conference and colloquium training can be expensive.

The quality of this type of training ranges from almost worthless to highly beneficial. Even the quality of courses offered by the same training institution will vary widely, depending upon the instructor. The value of seminars is largely a matter of how well the material presented can be related to improving the operations of your specific maintenance department. The best method of assuring this value is to thoroughly review the content of the proposed program, seek additional information as required, and be highly selective as to who attends. Sending five people can result in five times the loss of only sending one individual. If the one man reports back that the course was worthwhile, other individuals can be sent another time.

In summary, it can be stated that outside supervisory training usually has the drawbacks of being general in nature rather than being tailored to the specific requirements of the individual maintenance department. This drawback is one of the best arguments for having some type of in-plant training program.

TRAINING IN-PLANT

Despite the advantage of being able to tailor in-house training to specific departmental needs, in-plant training sessions for maintenance supervisors are often nothing more than a rain dance to give someone in the personnel department something to do. This failure lies with the maintenance manager who has not clearly stated the purpose or intent of the training. The program can become an end in itself, rather than a means to an end.

If a maintenance manager is going to have in-house training for his supervisors, he should constantly evaluate how well the program is improving his departmental performance.

In-plant training can be highly effective for the following purposes:

1. Learning how changes in personnel policies and procedures are to be implemented.

This type of training is particularly important when the work force is organized under labor union agreements. Supervision has to know the terms of each new contract and how they are to be administered. Otherwise there are bound to be violations, warranted grievances, and general friction. One maintenance department in Michigan reduced its grievances by 25 percent through a training program for maintenance supervision that covered each new labor union contract.

2. Learning the fiscal planning and accounting system of the parent organization.

This training can be invaluable. Maintenance supervisors should know how dollars are used in the planning and accounting system—how their budget and expenditures fit into the whole scheme of things. This knowledge will improve their effectiveness in planning, justifying, and controlling their expenditures. In one plant a course on this subject was the first step in developing a participative budgeting system. The new system reduced annual operating costs by seven percent in its first year of operation.

3. Training in work planning.

In one manufacturing plant, training in work planning for the maintenance foremen, combined with training in tools and materials methods improvement for the craftsmen, resulted in a 20 percent net reduction in maintenance costs from the preceding year.

4. Orientation to current operating plans of the parent organization.

A maintenance department is a part of an organization. If the supervisors in the maintenance department know the plans for the entire organization, they can understand the impact of what is planned and can determine what their section of the maintenance department can do to meet these objectives. This subject leads us to goal setting and a discussion of performance appraisal.

APPRAISING SUPERVISORS

There are three solid reasons for appraising supervisors. These are:

1. As a basis for rewards in the form of salary increases and promotions, or for discipline in the form of demotion or discharge.
2. As a guide for individualized development and training requirements.
3. As a means of implementing a management by objectives program.

The first two reasons have been the traditional reasons for performance appraisal. The third reason is somewhat new and has not been used extensively within maintenance departments.

USING MANAGEMENT BY OBJECTIVES

Applying management by objectives to performance appraisal essentially involves having each supervisor develop and be accountable for goals that are in consonance with the objectives of the maintenance department and the parent organization.

One maintenance manager puts it this way: "We used to have a performance appraisal method for supervisors that used attributes similar to the appraisal method for our craftsmen. But a few years ago I got enthusiastic about management by objectives. I feel that supervision should not be paid to figure out why they cannot meet goals. They should be paid for figuring out how to reach them. If we are objective oriented, we will seek ways to achieve the selected objectives. If we are problem oriented, we merely seek immediate solutions. Unfortunately, solutions obtained from a problem-oriented frame of reference usually solve symptoms rather than provide a lasting cure. Sometimes the solution may create more problems than it solves, because the focus of the solution is short range and based upon an immediate reaction rather than planned action. My supervisors are truly functioning when they are controlling methods, pace, and quality in conformance with plans. One way to develop these plans or objectives is to use the performance appraisal system."

Although the precise method of applying management by objectives may vary, the most successful systems apply all of the following considerations:

1. The first step should be a general briefing by the maintenance manager on what the goals and overall objectives of the total organization and the maintenance department are for a specified future period of time.

2. The individual supervisor then translates these goals into the impact they will have on his particular section. In turn, the supervisor determines what objectives his section will have to achieve to play its part in helping to achieve the overall goals.

3. The objectives of the supervisor are then drafted and reviewed with the maintenance manager.

   A critical responsibility of the maintenance manager is to properly evaluate the objectives during this review. Each objective should be clearly stated with respect to definition of task and method of measurement. The most specific method of stating goals is to quantify them. For example, to say that you are going to hire more apprentices is not nearly as specific as saying that you are going to increase the number of apprentices from three to five within a certain craft.

   The objectives should be compatible with overall goals, be practical, and actually represent a sufficient and attainable task for the supervisor during the measuring period. If the goals do not meet these criteria they should not be approved.

   After approval, the supervisor's goals are a statement of accountability for the period under consideration. Most systems use a twelve-month time period for performance.

4. The performance appraisal form used is a simple one. The specific goals are listed in one section of the form. At the end of the performance period, the results that were actually achieved for each objective are listed in another section. The supervisor's performance becomes obvious once the form has been completed at the end of the appraisal period.

5. An appraisal interview is then conducted between the maintenance manager and the supervisor to review the results.

One maintenance manager who has used the management by objectives method for several years feels that the appraisal interviews are now work centered with a constructive discussion of the results and ways to improve them. One section of the form says: "Here is what you said you would do," while the other section says: "Here is what I think you have done." There is no need for beating around the bush.

REWARDING AND DEVELOPING SUPERVISORS

If the supervisory appraisal system is not used to implement a management by objectives program, the first two reasons for having one still remain valid. If we get rid of all the jargon, appraisal is simply an attempt to think clearly about each supervisor's performance and future prospects against the background of his total work situation. Performance can be appraised succinctly by describing the best aspects of the supervisor's performance and identifying areas for improvement. Improvement may involve additional training. Future prospects can be reviewed constructively when the appraiser attempts to identify jobs the supervisor could perform successfully if given the opportunity.

16 Implementing Successful Maintenance Apprenticeship Programs

According to figures released by the U.S. Department of Labor, there were over 17,000 active registered apprentices in the United States in the 1970s. These apprentices were indentured in a wide variety of trades. However, the use of apprenticeship programs in maintenance departments is not extensive, probably because of a lack of long-range planning on the part of maintenance managers in performing the staffing function. There is a failure to recognize the need to provide skilled craftsmen on a continuing basis to execute the operations assigned to their department in achieving organizational goals.

This chapter covers facets of successful programs that can be used in developing a formal apprenticeship program in your maintenance department.

Planned staffing can definitely enhance departmental performance. Proper execution of staffing, particularly in filling the need for skilled journeymen, is inherent in an apprenticeship program. Apprentices are hired on the basis of known needs for future journeymen, and the training requirements for an apprentice are thoroughly formalized. Those maintenance departments that do have apprenticeship programs have resolved their problems in the staffing of skilled journeymen. It is also important to note that a large number of journeymen who have been trained in an apprenticeship program have moved into supervisory and management positions.

One company that has an apprenticeship program increased productivity in maintenance and operations by over 15 percent. The increase was attributable not only to an increase in the skill levels within the department, but also to the fact that the work processes that had to be assigned to the apprentices as part of their training forced the department to do a better job of planning for the total workload. Although there are no "before and after" studies available, it is generally conceded that apprenticeship programs also reduce labor turnover.

WHAT APPRENTICESHIP IS

Apprenticeship is a system of "learning while earning" and "learning by doing." It combines training on the job with related technical instruction at school. It operates under a written program of training standards agreed to by employees, journeymen, and apprentices, whereby a young person gains from the skilled journeyman his skill and know-how, and in turn becomes an important part of his trade. In fact, apprenticeship has proven to be the only universal, time-tested way to provide all-round skilled craftsmen.

ADVANTAGES OF AN APPRENTICESHIP PROGRAM

Depending upon who you talk to, the advantages of having an apprenticeship program will vary. The following five items should all be considered in evaluating whether or not you want to start a program in your maintenance department.

1. It upgrades the skills and competency of maintenance and operations personnel, and has a direct, positive impact on departmental productivity and provides for employee career development.
2. It provides a direct career path for unskilled workers in entry level positions both within the maintenance department and other departments of the organization.
3. It facilitates the hiring of young people. Without such a program the normal course of action is to hire only experienced journeymen. Experienced journeymen are in short supply, and the older they are the fewer productive years they have left in which to be totally effective.
4. It establishes a continuing source for the replenishment of qualified journeymen. At present, many journeymen in the labor market have not completed formal apprenticeship programs. Without such a certification, the skills and competency of a journeyman are not readily determinable, and the probability of hiring fully qualified journeymen is lessened. When you have trained the journeyman as an apprentice, you know his qualifications.
5. It promotes an association of mutual interest and respect between the company and the employee, which results in greater job satisfaction and reduced turnover.

HAVING A STATE-APPROVED PROGRAM

There are some definite advantages in having an apprenticeship program approved by the state in which your program resides. The major advantage is that state agencies provide coordination, give service and supervision, and insure the maintenance of sound standards designed to protect all participants in the program.

1. The state provides an apprenticeship consultant from its apprenticeship agency who will assist in maintaining the program. The standards established for the program are in conformance with the standards used throughout the state.
2. The local public schools provide a consultant to the program and the company gets the support of local public schools in determining the technical classes of supplemental instruction that are to be attended by apprentices.
3. The State Department of Employment or Human Resources Development provides the services of screening apprentice applicants and providing access to unemployed apprentices.
4. A state-approved program qualifies apprentices who are eligible veterans for training allowances under the Veterans Assistant Act, which creates an

additional incentive for enrollment and continuance of training as an apprentice.

THE STATE LAWS ON APPRENTICESHIP

Most states have an Apprenticeship Labor Standards Act. For those that do not have a state law, a program is administered by the U.S. Department of Labor's Bureau of Apprenticeship and Training.

With a few exceptions, a state law is normally voluntary. The rules and regulations set up by the state apprenticeship agency are purposely flexible so as to meet changing local situations, the only objective being to insure high standards for the training of skilled men.

Normally, a state apprenticeship law provides that:

1. An apprentice must be at least of a certain age (usually 16 years).
2. He enters into a written agreement with an employer or his agent.
3. His training period shall be not less than 2,000 hours (roughly one year).
4. He shall attend classes of related supplemental instruction.
5. Upon successful completion of his apprenticeship, after having met the requirements of his agreement, the apprentice will be awarded a certificate of completion by the state, upon recommendation by his employer and the apprenticeship committee.

MANAGING AN APPRENTICESHIP PROGRAM

Usually, an apprenticeship program is administered either by a Joint Apprenticeship Committee (JAC) or a Unilateral Apprenticeship Committee (UAC).

A JAC consists of representatives of the cognizant trade union for the apprenticed trade and representatives of the companies who employ the apprentices.

A UAC consists of representatives of only an individual employer who establishes a program for strictly his own company. An enterprise does not have to be large to have a UAC. In 1972 an individual proprietorship employing only three people established its own UAC in the state of California for an apprenticeship in saddle making. By the same token, a maintenance department does not have to be a very large operation in order to have an apprenticeship program.

A variation of the Unilateral Apprenticeship Committee is a local program sponsored by an association of employers without the participation of an employee labor union. This type of a program can be beneficial when a given employer does not have the resources to properly train an apprentice in a certain work process. The apprentice can be temporarily transferred to another employer in the association in order to acquire the training.

The apprenticeship committee is the governing body of an apprenticeship program. To assist the committee, the state provides an apprenticeship consultant from its apprenticeship agency and a representative from the local public schools. The committee sets the standards, rules, and regulations to ensure that apprentices are hired and trained in the trades included in the program.

MAINTENANCE TRADES INCLUDED IN A PROGRAM

An apprenticeship program at a company may include one or more maintenance trades. Some examples of apprenticeship trades are as follows:

Maintenance Trade	Length of Apprenticeship
1. Electrician	4 years
2. Air Conditioning Mechanic	4 years
3. Mechanic Machinist	4 years
4. Plumber	4 years
5. Sheet Metal Mechanic	4 years
6. Carpenter	4 years
7. Office Equipment Mechanic	4 years
8. Painter	3 years
9. Landscape Gardener	2 years

As an example, the details on the specific work processes and approximate hours of training for the apprenticeship for the occupational trade of maintenance electrician are contained in Figure 16-1.

In addition to the nine trades listed above, one company also has a four-year apprenticeship program for electronic electricians to support the maintenance of their numerical tape-controlled machine tools in the manufacturing departments. Apprenticeship can be made to keep in step with advancements in technology and is not limited to the more classical trades associated with the construction industry, such as plumbing, painting, carpentry, or sheet metal work.

APPRENTICE WAGES AND BENEFITS

The wage of an apprentice is set up as a percentage of that earned by a journeyman. Assuming satisfactory performance, the apprentice's wage is increased every six months during the period of his apprenticeship. Upon completion of his apprenticeship, he receives the maximum pay of the wage grade for a journeyman in the trade.

```
┌─────────────────────────────────────────────────────────────────┐
│                   MAINTENANCE ELECTRICIAN                         │
│                        APPRENTICE                                 │
│                                                                   │
│                                             APPROXIMATE           │
│            WORK PROCESSES                      HOURS              │
│                                                                   │
│  A. Familiarize the apprentice with shop practices, operation     │
│     and use of equipment, hand and power tools, and               │
│     materials.                                       200          │
│                                                                   │
│  B. Safety, work practice, education and understanding of         │
│     State Safety Orders.                             100          │
│                                                                   │
│  C. Fixtures (interior and exterior lighting equipment, light     │
│     service).                                      1,000          │
│                                                                   │
│  D. Controls (electronic controlled equipment, AC-DC             │
│     motors, power service, AC-DC generators, heating equip-      │
│     ment, protective devices, switch boards, switch gear,        │
│     transformers).                                 1,800          │
│                                                                   │
│  E. Alarm systems (electronic controlled equipment, instru-      │
│     ment and process controlled equipment, public address        │
│     systems, signal systems).                        700          │
│                                                                   │
│  F. Wire, conduit, and switch installation, remodeling and       │
│     construction.                                  1,200          │
│                                                                   │
│  G. Electric motor repairs and maintenance.        1,200          │
│                                                                   │
│  H. AC controls maintenance and repair.            1,200          │
│                                                                   │
│  I. Blueprint reading, National Electrical Code, instrument       │
│     measuring, labor-material estimating and layout. Theory       │
│     (basic electronics). AC-DC three phase power distribu-        │
│     tion.                                            600          │
│                                           TOTAL    8,000          │
│                                                                   │
│                        FIGURE 16-1                                │
└─────────────────────────────────────────────────────────────────┘
```

WORK PROCESSES FOR A MAINTENANCE ELECTRICIAN APPRENTICE

The company-paid benefits for apprentices are normally identical to the benefits provided other permanent, full-time company employees. In addition to these benefits there are, of course, the statutory benefits such as Social Security, State Disability and Unemployment Insurance and Workman's Compensation. Some companies also offer the optional benefits of additional employee and dependent life insurance on an employee-paid basis. Long-term disability insurance and a stock savings plan are also sometimes offered, and may be partially employee paid and company supplemented.

For those individuals who are not veterans receiving training allowances under the Veterans Assistance Act, the company will usually pay 100 percent of the books and tuition charges for the formal technical classroom training taken by the apprentice. Reimbursement is normally made upon successful completion of the course of instruction.

THE OPPORTUNITIES FOR THE APPRENTICE

Ability to earn is the prime factor in determining a worker's standard of living—the greater the ability, the higher the standard. Among workers, skilled craftsmen have the highest earning capacity. Apprenticeship makes these craftsmen. The advantages to an individual who enters an apprenticeship program provided by a state are:

1. A steady guaranteed pay scale while in training, with step-by-step increases at regular intervals, backed by written agreement with his employer and the state
2. An organized training program, specifically designed to rotate him through all the job processes of his chosen trade, in order to insure well-rounded, thorough training while working
3. Special supplemental schooling in subjects related to the trade but usually not available in the shop
4. Qualification of apprentices who are eligible veterans for training allowances under the Veteran's Assistance and Readjustment Act
5. A certificate from the state, upon completion, as proof of satisfactory training and competence. This certification qualifies him to be eligible to join any trade-related labor union in the state as a recognized journeyman
6. The opportunity to draw top wages as a skilled tradesman

Because of these opportunities, the recruiting for an apprenticeship program is not a difficult task. There are usually plenty of qualified applicants.

GETTING APPRENTICES

It must be recognized that not every young person is qualified to become a craftsman. Rigid methods of selection must be used to obtain only the finest young candidates for future journeymanship since only competent craftsmen skilled in all branches of their trades can form the basis of an effective maintenance department.

Therefore, after an apprenticeship program is established, the first step must be a careful selection of the people from which journeymen will be evolved.

Getting apprentices into a program is normally accomplished in three steps. These are:

1. Recruitment of applicants
2. Creation of a pool of eligibles
3. Selection from the pool

RECRUITING

For a state-approved apprenticeship program there are many resources available for the recruitment of applicants. Some examples are:

1. The State Apprenticeship Agency
2. The state supervisor of the Bureau of Apprenticeship for the U.S. Department of Labor
3. The local public school district
4. The State Employment Agency

The State Apprenticeship Agency and the U.S. Department of Labor can sometimes provide applicants who have received previous apprenticeship training in the trade either within the state or in some other state. The local public school districts through the County Superintendent of Schools and Community College Districts can provide applicants who have had vocational training. The State Employment Agency does not only provide potential apprentices, but will often administer any prescribed aptitude tests to candidates obtained from all sources.

Regardless of the recruitment source, all applicants should be required to fill in an application blank for future reference in contacting the applicant.

CREATING A POOL

A pool of eligible candidates is created from those applicants who are able to meet the qualification standards. The standards normally applied are:

1. Age
2. Physical Ability
3. Educational Attainment
4. Aptitude Tests
5. Oral Interviews

AGE, PHYSICAL ABILITY, AND EDUCATION

The first three of the standards listed above are rather easily applied. It is relatively simple to ascertain if a candidate is within the prescribed age limitations. The physical requirements of the program can be met by requiring a declaration of any physical impairments by the candidate on the written application form. An examination by a qualified physician should normally be reserved until applicants have been selected from the pool. Considerable expense can be incurred if all applicants receive a physical examination, rather than just those few who are ultimately selected for hiring. Educational attainment (e.g., high school graduation or general education development test grade) is also readily ascertained by the evidence of a diploma or certified test score.

APTITUDE TESTS

Aptitude tests require more effort in being applied as a qualification standard because the particular test must be shown to be both valid and reliable. Reliability is measured by consistency in the measurement of whatever aptitude the test is endeavoring to measure. Validity is determined by whether the test actually measures whatever aptitude it is supposed to measure. Many tests have recently fallen into disrepute because they have been shown to lack validity when used to test members of minority groups. They were discovered to be valid when used for the testing of middle-class urban whites, but lacking validity in testing these same aptitudes in certain racial minority groups.

Any qualification standard for admission to the pool consisting of aptitude test scores should be directly related to job performance. Relationship to job performance should be shown by significant statistical and practical relationships between the score on the aptitude tests and the score required for admission to the pool and performance in the apprenticeship program. If the relationship cannot be demonstrated, the tests should not be used as a qualification standard.

ORAL INTERVIEWS

The tool used most often in personnel administration for the assessment of applicants for job openings is the interview. The interview is used extensively in determining which applicants should be placed in a pool of eligibles and for the ranking of eligibles in a pool of apprenticeship applicants.

When the oral interview is used, it should be judiciously structured to be as objective as possible. Toward this end the following should be applied:

1. Use a standardized set of questions to assure that all applicants are asked the same questions.
2. Record the general nature of the applicant's answers on the questionnaire form.
3. Record a conclusion for each specific factor used in the questionnaire.
4. Record a brief summary of each interview on the questionnaire form.

In an effort to further reduce the subjectivity inherent in an interview, a point value can be assigned to each specific factor to be measured in the interview. In turn, gradations of point values can be established within the total point value assigned to a specific factor.

As an example of the application of these methods, let us analyze the oral interview form appearing in Figure 16-2 on the next four pages.

FIGURE 16-2

APPRENTICE APPLICANT ORAL INTERVIEW QUESTIONNAIRE

Occupation: _____

Name of Applicant: _____
                    Last   First   Middle

Date of Interview: _____

Grade Summary:

Factor	Rating	Numerical Grade
1. Attitude	_____	_____
2. Confidence	_____	_____
3. Motivation	_____	_____
4. Stability	_____	_____
5. Oral Response	_____	_____

Total Grade   Pass ☐   Fail ☐

NOTE: A minimum passing grade on this oral interview is 70%. Each of the five factors is equally weighted, with a maximum of 20 points. Descriptive ratings within a factor are assigned the following numeric points:

Outstanding = 20   Acceptable = 15   Mediocre = 10   Poor = 5

Summary Conclusions:

_____

_____

_____

_____

_____

_____

_____

_____

_____

_____

Interviewers:

Signature: _____   Date: _____

Name (print): _____
              Last   First   Middle

Signature: _____   Date: _____

Name (print): _____
              Last   First   Middle

Signature: _____   Date: _____

Name (print): _____
              Last   First   Middle

Page 1 of 4

Name of Applicant: _____

Last   First   Middle

ATTITUDE

Rating

Outstanding ☐   Acceptable ☐   Mediocre ☐   Poor ☐

Questions

1. Why do you want to become an apprentice?

Summary of Reply

_____

_____

_____

_____

2. Do you think that you can apply yourself both on-the-job and in formal classroom instruction courses required during the length of this apprenticeship? (State length of apprenticeship.)

Summary of Reply

_____

_____

_____

_____

CONFIDENCE

Rating

Outstanding ☐   Acceptable ☐   Mediocre ☐   Poor ☐

Questions

1. Do you feel that you can pass the formal classroom instruction courses required as an apprentice? (State courses that will be required.)

Summary of Reply

_____

_____

_____

_____

APPRENTICE APPLICANT ORAL INTERVIEW QUESTIONNAIRE *(cont'd)*

Name of Applicant: _____
                                   Last   First   Middle

   2. Are you confident that you can learn all of the work processes required for this occupational trade? (Review work processes per standards.)

     Summary of Reply

_____

_____

_____

MOTIVATION

Rating

Outstanding ☐   Acceptable ☐   Mediocre ☐   Poor ☐

Questions

1. Why do you desire to learn this occupational trade?

    Summary of Reply

_____

_____

_____

_____

2. What are your job career goals after becoming a journeyman?

    Summary of Reply

_____

_____

_____

_____

STABILITY

Rating

Outstanding ☐   Acceptable ☐   Mediocre ☐   Poor ☐

Questions

1. What was your attendance record in high school?

    Summary of Reply

_____

_____

APPRENTICE APPLICANT ORAL INTERVIEW QUESTIONNAIRE *(cont'd)*

Name of Applicant: _____
                           Last   First   Middle

2. Did you graduate from high school?   Yes ☐   No ☐

   Summary of Reply

   _____

   _____

3. What type of work are you doing now and what has been your longest period of employment?

   Summary of Reply

   _____

   _____

   _____

   _____

4. If you have a driver's license, has it ever been revoked?

   Yes ☐   No ☐

   Summary of Reply

   _____

   _____

ORAL RESPONSE

Rating

Outstanding ☐   Acceptable ☐   Mediocre ☐   Poor ☐

The grading for this item is determined by how well the applicant responded to the questions asked regarding attitude, confidence, motivation, and stability (i.e., were the answers clear, concise, and well-conceived, demonstrating his ability to express himself verbally?)

Basis for Rating Assigned

_____

_____

_____

_____

_____

_____

_____

This form is used by the maintenance department of a Southern California firm. The first page is a summary sheet. It contains the standard data with respect to the apprentice trade, the names of the applicant and interviewers, the date of the interview, and so on. The lower center portion of the page contains space for a summary of the conclusions of the interviewers.

The top center portion of the form is used as a grade summary. The interview covers attitude, confidence, motivation, stability, and oral response. For each factor a space is provided for a descriptive rating and an assigned numerical grade for the rating.

The descriptive ratings are transferred from subsequent completed pages in the questionnaire and recorded in the grade summary portion of the first page. Note that each of the five factors is equally weighted, with a maximum of twenty points. Thus, the maximum score is 100 (five factors times a maximum point value of twenty). Within each factor there are four descriptive gradation ratings with assigned numerical values. The gradations range from outstanding, with an assigned value of twenty points, to poor, with an assigned value of five points. Assuming that a candidate was assigned the rating of acceptable for all five factors, his score would equal seventy-five points (five factors times fifteen points per factor). Thus, he would have achieved the minimum passing score of seventy.

In the subsequent pages of the form the first four factors covered require specific questions to be asked of the applicant with a space for recording the general nature of the answer. Note that some of these questions are dichotomous. That is, they provide for strictly a yes or no answer. Other questions are of the open-end type, permitting the applicant to respond in any manner he desires. A descriptive rating is recorded for each factor. These descriptive ratings are then transferred to the grade summary portion on the first page of the questionnaire and a numerical grade is calculated.

The fifth factor, oral response, is evaluated by the applicant's response to the specific questions asked about the first four factors. In questionnaire design, this method is sometimes called the indirect derivation technique. When it is used, care should be taken to assure that the interviewer has received some training in interviewing. If used by a novice, it can be criticized as being too subjective.

SELECTING ELIGIBLE APPLICANTS

One of two methods is usually used to select applicants from a pool of eligibles. These are:

- Random Selection
- Rank Selection

RANDOM SELECTION

The classic method of random selection is to draw names from a hat or draw bingo numbers to which the names of applicants have been previously assigned.

If the random selection method is used, certain steps should be followed to avoid any charges of collusion. The drawing should be supervised by an impartial person or persons not associated with the administration of the apprenticeship program. The time and place of the drawing should be announced and subsequently open to attendance by all applicants and the public. The number of applicants selected should exceed the number of positions available so that in the event a selected candidate fails to pass the physical examination, an alternate is immediately available, thereby eliminating the need for a second drawing.

The main disadvantage in using the random selection method is that you may not get the best-qualified applicants. Obviously, not all eligible candidates in the pool are of equal quality. Some are bound to be better than others. Using random selection, you may end up drawing some of the less-qualified applicants.

RANK SELECTION

The rank selection method involves the ranking of each eligible applicant in relation to all other candidates in the pool. This ranking is usually done on the basis of the total point grade assigned in the oral interview. Aptitude test scores are not normally used because the test results are sometimes reported only as a "pass" or "fail," rather than as specific score results.

After the eligible applicants have been ranked by their oral interview score, there are instances when one or more candidates have the same score. The order of ranking of individuals with tie scores may then be accomplished on the basis of seniority in making written application to the program.

After ranking, hiring is done following sequential order from the top of the list. If a candidate fails to pass the physical examination, the next ranked candidate is taken as a replacement.

TRAINING APPRENTICES

The training of an apprentice involves the following two dimensions:

1. Classroom Instruction.
2. Training in Specific Work Processes.

CLASSROOM INSTRUCTION

Formal classroom instruction is a required part of an apprentice's training. A specific set of technical courses that are to be taken during the length of the apprenticeship must be developed. Where there is a large number of apprentices in a given craft, the local school district will often establish specific classes for a given program or even a specific curriculum of classes that is given to fulfill the entire apprenticeship program needs. In other instances, such as for a specialized trade, it may be necessary to develop a special curriculum using those courses that are available in local colleges and vocational schools and also maintenance training courses given by equipment manufacturers. The apprentice should be required to demonstrate satisfactory class attendance and academic achievement or he should be dropped from the program.

Figure 16-3 is a representative listing of the curriculum of courses for a maintenance electrician used by one company. The course contents are described in Figure 16-4. In developing a curriculum for a given apprenticeship trade, the representative of the local school district can lend a tremendous amount of help, both in locating courses that are currently available and in establishing new courses to meet the needs of the program.

APPRENTICESHIP PROGRAM
Craft: Maintenance Electrician
Length of Apprenticeship: Four Years

FORMAL CLASSROOM INSTRUCTION REQUIREMENTS

General Description of Instruction

   Tools and equipment; electrical principles and application to basic AC-DC circuitry; motors, generators, controls, transformers; electrical codes and ordinances; related mathematics and electrical blueprint reading; safety practices.

Specific Courses

Title	Hours of Credit
Electrical Theory	54
Electrical Tools and Materials	54
Electrical Estimating and Blueprint Reading	54
Electrical Codes and Ordinances	108
Electrical Motor Controls	108
Motor Control Equipment Maintenance	54
Total	432

FIGURE 16-3

FORMAL CLASSROOM INSTRUCTION CURRICULUM

Surveillance of the apprentice's attendance at prescribed classes of formal instruction should be effected. One method of providing visibility is a monthly attendance certification record from the class instructor. Figure 16-5 is an example of such an attendance record. Note that the form also provides for the instructor's rating with respect to the classroom performance of the apprentice, which provides an early rating before the end of a semester.

APPRENTICESHIP PROGRAM
Craft: Maintenance Electrician

FORMAL CLASSROOM INSTRUCTION REQUIREMENTS
DESCRIPTION OF TYPICAL COURSE CONTENT

1. Electrical Theory—One Semester
   This course gives technical and related information to supplement the practical training received while working at the electrical trade as an indentured apprentice. It covers electron theory, electrical units, application of Ohm's Law, electrical power, instruments, resistance of wire, with extensive mathematical applications.

2. Electrical Tools and Materials—One Semester
   This is a shop course in the selection and use of the proper tools for assembling electrical components and devices.

3. Electrical Estimating and Blueprint Reading—One Semester
   This course acquaints the student with the fundamental concepts involved in electrical estimating and blueprint reading, such as material takeoff and extension, determining and assigning labor units, proper interpretation of specifications and blueprints, and job variables and codes as they affect the ultimate price. Students work with actual prints and job specifications, make material lists, and assign labor units to determine the estimated price for electrical installations.

4. Electrical Codes and Ordinances—Two Semesters
   These courses give instruction in basic employment information. Safe job practices are stressed. In addition, they supplement on-the-job training with instruction in identification of electrical materials. Training is given in rigging methods, ropes, and knots. Emphasis is placed on tools used for mechanical advantage and their safe use on electrical installations.

5. Electric Motor Controls—Two Semesters
   These courses provide instruction in basic motor control fundamentals including the basic aspects of control, review of magnetic principles of DC and AC motors, types of motors, and motor selection fundamentals, discussion of definitions for controller components and symbols, familiarization of N.E.M.A. standards, and review of one-line, wiring, and schematic diagrams (the schematic diagram is emphasized). The magnetic controller is studied in detail. The selection and application of DC and AC controllers, with emphasis on the AC devices, are covered. Manual, magnetic, across-the-line starters, and most forms of reduced voltage starters are discussed, including the autotransformer, primary resistor, star-delta, partwinding and wound rotor-type reduced voltage starters. Synchronous and multispeed starters, the many methods of decelerating or braking, and static components are discussed, with emphasis on the schematic diagram in every case.

6. Motor Control Equipment Maintenance—One Semester
   This is a course in which students draw schematic diagrams and connect and test two-wire and three-wire control devices. Students learn how to identify, tag, and connect nine-lead wye and delta three-phase motors, series shunt motors, and compound DC motors, and how to reverse all types of motors.

FIGURE 16-4

DESCRIPTION OF COURSE CONTENT

MONTHLY CLASSROOM INSTRUCTION RECORD

Apprentice Name: _____

Trade: _____

Name of School: _____

Course No. and Title: _____

Name of Instructor: _____

ATTENDANCE RECORD

Month of: _____ , 19 ____

Date	No. of Hours		Date	No. of Hours
_____	_____		_____	_____
_____	_____		_____	_____
_____	_____		_____	_____
_____	_____		_____	_____
_____	_____		_____	_____
_____	_____		_____	_____
_____	_____		_____	_____
_____	_____		_____	_____
_____	_____		_____	_____
_____	_____		_____	_____

VERIFICATION OF ATTENDANCE

Instructor's Signature: _____ Date: _____

Instructor's Rating: (Please check the applicable rating block and make a brief comment regarding the above-named apprentice's apparent academic progress, interest and attitude with respect to the course.)

Rating: Outstanding ☐   Acceptable ☐   Mediocre ☐   Poor ☐

Comments:_____

_____

_____

_____

_____

FIGURE 16-5

SAMPLE CLASSROOM ATTENDANCE RECORD

In order to emphasize the importance of formal classroom instruction, some apprenticeship committees utilize formalized disciplinary measures for unjustifiable absences from class. One program invokes the issuance of a warning notice after the second absence, three day's suspension from work for a third absence, and cancellation of the apprenticeship agreement for a fourth absence during a semester course.

TRAINING IN WORK PROCESSES

Training in the work processes involves three people: the foreman, the journeyman, and the apprentice.

The Foreman

The foreman must plan the work schedule of an apprentice in such a manner that he will be assured of receiving adequate hours of training in each of the work processes. Otherwise, the apprentice will not be able to complete the program in the time prescribed.

The foreman must also be responsible for the assignment of a qualified journeyman to work with the apprentice during the performance of each of the work processes. He must assure that the journeyman assigned is properly instructing the apprentice.

The Journeyman

Several journeymen may be involved in the training of an apprentice due to vacations, changes in shifts worked, or the selection of a journeyman for training in a specific work process based upon the fact that the particular journeyman has the best ability to teach the skills associated with the work process.

The journeyman is an instructor. He must take the time to explain to the apprentice how the work is done. A hard-working journeyman is sometimes a poor teacher since he is more concerned with getting the job done than with training the apprentice. It should also be recognized that the time a journeyman takes to instruct an apprentice will decrease his productivity to some degree.

The Apprentice

The apprentice is the third party involved in the work process training. He should know that he is there to learn certain skills by working on the job. He must be prepared to ask questions when he does not understand what is being explained to him.

The apprentice is required to keep track of the hours expended on each work process. Figure 16-6 is an example of a work process recording sheet. Sheets are normally turned in monthly for posting to the apprentice's progress record so that his work assignments can be tracked to assure that he is getting training in all the work processes and has not exceeded the hours required for a given work process. Figure 16-7 is an example of a progressive training sheet covering a six-month period. Note

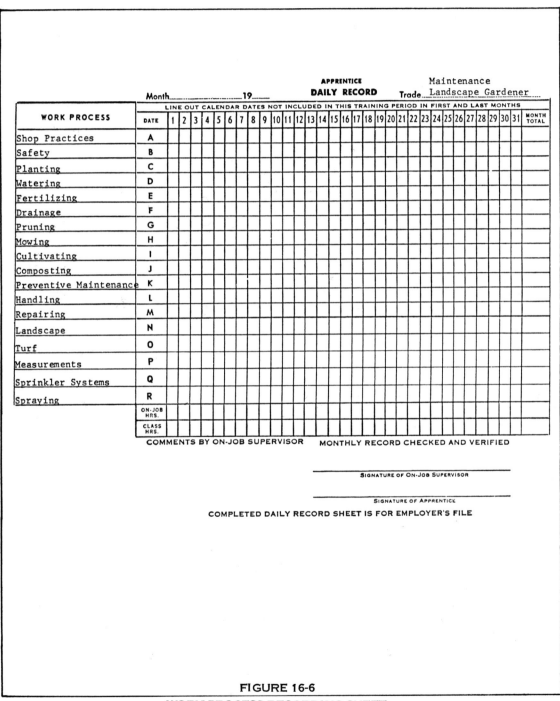

WORK PROCESS	DATE	1	2	3	4	5	6	7	8	9	10	11	12	13	14	15	16	17	18	19	20	21	22	23	24	25	26	27	28	29	30	31	MONTH TOTAL
Shop Practices	A																																
Safety	B																																
Planting	C																																
Watering	D																																
Fertilizing	E																																
Drainage	F																																
Pruning	G																																
Mowing	H																																
Cultivating	I																																
Composting	J																																
Preventive Maintenance	K																																
Handling	L																																
Repairing	M																																
Landscape	N																																
Turf	O																																
Measurements	P																																
Sprinkler Systems	Q																																
Spraying	R																																
ON-JOB HRS.																																	
CLASS HRS.																																	

Month_____19____

APPRENTICE DAILY RECORD   Trade Maintenance Landscape Gardener

LINE OUT CALENDAR DATES NOT INCLUDED IN THIS TRAINING PERIOD IN FIRST AND LAST MONTHS

COMMENTS BY ON-JOB SUPERVISOR   MONTHLY RECORD CHECKED AND VERIFIED

SIGNATURE OF ON-JOB SUPERVISOR

SIGNATURE OF APPRENTICE

COMPLETED DAILY RECORD SHEET IS FOR EMPLOYER'S FILE

FIGURE 16-6

WORK PROCESS RECORDING SHEET

APPRENTICESHIP PROGRAM
WORK PROCESS SUMMARY SHEET

TRADE _____

APPRENTICE NAME _____

WORK PROCESS	PROGRAM REQUIRED	PREVIOUS CREDIT	REMAINING REQUIR'NTS.	COMPLETED	BALANCE REQUIRED	COMPLETED	BALANCE REQUIRED	COMPLETED	BALANCE REQUIRED	COMPLETED	BALANCE REQUIRED	COMPLETED	BALANCE REQUIRED	COMPLETED	BALANCE REQUIRED	REMARKS
A																
B																
C																
D																
E																
F																
G																
H																
I																
J																
K																
L																
M																
N																
O																
P																
Q																
R																

FIGURE 16-7

PROGRESSIVE TRAINING RECORD SHEET

212

that the sheet includes seven months. The training period commenced sometime during the month of July and must carry over into January to provide for a full six-month period.

Progress Review Sessions

Periodic review sessions among the foreman, the journeyman giving the training, and the apprentice should be a definite part of the program. These review sessions facilitate surveillance of the training by the foreman and provide early identification of any conflicts between the journeyman and the apprentice. During the initial steps of the apprenticeship, these review sessions should be held quarterly or more frequently to assure that things are starting off properly. There have been instances where the first progress review session was not held until the end of the first six months when the apprentice was to receive his evaluation for advancement to the second step in the program. Such a long wait penalizes some of the apprentices. Areas in which there is need for improvement are not identified soon enough for them to take corrective action in time to assure their advancement to the next step. One such review revealed a marked personality conflict between the apprentice and the journeyman assigned to teach him the work processes. The foreman did not recommend the apprentice's advancement because the journeyman claimed that the apprentice may have logged the hours, but "had a lousy attitude and had not really learned anything." The apprentice was reassigned to another journeyman and subsequently performed successfully. However, he was delayed six months in the program because the initial review session was held too late to provide for an early identification and rectification of the situation.

Review sessions also serve another purpose. At some point in the program there is usually a loss of enthusiasm by the apprentice. The letdown seems to occur most often around the end of the first year. A review session seems to provide reassurance to the apprentice and gets him going again.

SECTION IV–CONTROL

Food for thought:

- *There is a right time for everything.*
- *Laziness lets the roof leak and soon the rafters begin to rot.*
- *You are a poor specimen if you can't stand the pressure of adversity.*

POINTS TO CONSIDER

Management control consists of assuring that events are conforming to plans. This task is achieved by measuring and correcting the performance of subordinates in order to make sure that what is done is what is intended. To achieve this goal an adequate control system must disclose where deviations are occurring, who is responsible for them, and what should be done about them.

This section of the book contains six chapters on the management function of control. The first chapter provides an overview of the control function. The next three chapters cover proven work reception, work authorization, and performance analysis techniques used to achieve control. The next chapter discusses successful applications of automated information systems to support the control function. The last chapter tells about the necessary steps to be used in developing effective automated systems.

17 Establishing Maintenance Control Systems That Work

The purpose of control is to make sure that people are doing what has been planned. This chapter is the first of six chapters on control. It is an important chapter because it provides some insight into what control is all about, defines the essential elements of a control system, and covers proven control techniques.

Some maintenance managers find that the easiest way to explain control is to say that it is one side of a coin that has planning on the other side. This explanation may be an oversimplification so far as some academicinas are concerned, but the illustration is essentially correct in describing the interrelationship between the two management functions. Plans are meaningless if there is no way to determine the extent and quality of their implementation through controls. Likewise, without goals or plans there are no criteria against which events can be compared for conformance.

Certain management theorists prefer to use the word "measurement" in lieu of the word "control" to identify this management function because, in their opinion, control is a dirty word. It has the connotation of limiting freedom by exercising a restraining influence over people. The fact that these theorists seem to overlook is that control is best exercised by creating a state of mind that recognizes a commitment to achieve certain plans or objectives. Control should result from that commitment rather than from a restraining influence on people. The process of control is not merely the use of measurement, a budget, a variance analysis, or a pile of reports. These are the paraphernalia that are supposed to help in achieving the goals or plans. They are merely means towards the end of achieving control.

Plans are expressed in units of measure. Control is facilitated by measuring performance in these same units. Control, however, is not achievable strictly through the application of mathematical formulas or measurement. Quantification does add a preciseness to planning and control, but just because status is reported in terms of numbers does not assure that events are completely conforming to plans. There are other essential elements to a successful control system.

HAVING A CONTROL SYSTEM

To achieve effective execution of management control for a given task or mission requires a system that contains certain essential elements. These elements are:

- Plans and Procedures
- Measuring
- Reporting
- Reviewing
- Decision Making

HAVING PLANS AND PROCEDURES

As previously stated, there can be no control unless there are plans and procedures for what is to be accomplished. Although the type of plan or procedure used will vary according to the task over which control is to be executed, there has to be some concept of what is to be done and the procedures that are to be used in order to accomplish control. Plans define what is to be done, while procedures assign responsibilities within the organization structure for getting the plans accomplished. Chapter 18 deals with work reception procedures to facilitate control. Chapter 19 discusses work authorization procedures as a means of achieving control.

MEASURING

To determine if plans are being executed properly, it is necessary to have some means of measurement. The most effective means generally rely on some method of quantification.

Schedules are set up to plan tasks in relationship to time. Control is used to measure performance by determining if the tasks are accomplished within the stipulated time periods.

A budget is set up to plan tasks in relationship to time and dollars. Control is used to measure performance by determining if the tasks are accomplished within the stipulated budget time period and dollar amounts.

Other methods of quantifying plans previously discussed with respect to planning are headcount, man-hours at straight- and premium-time pay, and equivalent heads. All of these methods are types of quantified measurement.

REPORTING

After developing plans and procedures and methods of measuring performance against plans, the next essential step is to have a reporting system to reveal what is being accomplished.

Depending on what activity is being controlled, the reporting method may be verbal, manual (handwritten or typewritten), or automated. The application and development of automated information systems is further discussed in chapters 21 and 22.

It should be recognized that reports do not necessarily have to be prepared in the maintenance department. For example, the most widely used method for reporting on budget performance is the cost accounting system which contains dollar expenditures. The timekeeping system may be used to prepare written reports on in-house labor hours both at straight- and premium-time pay. Verbal reports may also come from personnel who are not part of the maintenance department. Most telephone calls to a trouble call desk are made by people in other departments.

A project scheduling system that provides for the reporting of actual completion dates of milestones within the schedule serves both as a planning and control tool.

Figure 17-1 is a sample format. The Estimated Date blocks for each activity are used to enter the estimated or scheduled date. The Actual Date blocks are used to post the actual dates when actions were taken.

The use of a format similar to Figure 17-1 to schedule and maintain status control on construction projects helped one maintenance department improve its delinquency or behind-schedule completions by 20 percent.

PROJECT SCHEDULE AND PERFORMANCE

Project Number: _____

Description: _____

Activity		Estimated Date	Actual Date
Design	Start		
	Complete		
Preliminary Cost Estimate	Start		
	Complete		
Design Review and Changes	Start		
	Complete		
Final Cost Estimate	Start		
	Complete		
Signature Approvals	Start		
	Complete		
Bidding	Out		
	Opening		
	Award		
Contractor Work	Start		
	Complete		
Final Fit-Up	Start		
	Complete		
Occupancy	Start		
	Complete		

FIGURE 17-1

SAMPLE PROJECT SCHEDULE

REVIEWING REPORTED DATA

Plans, methods of measurement, and reporting methods are of little value unless the performance data is reviewed to identify deviations from the plans. If budgets are being underrun or overrun, or if schedules are not being adhered to, little can be done to rectify the situations unless the specific deviations are identified for management action. The reported data needs to be reviewed to determine where deviations from plans are occurring. Normally, a review of written data is accomplished by staff personnel in the finance or planning and control section of the maintenance department. Regardless of which staff personnel accomplish the task, the detailed written data should be summarized into meaningful management reports and the specific deviations identified for supervisory action.

The technique of exception reporting relies on reporting only on those jobs or tasks where deviations are occurring. The bottlenecks are identified and management efforts can be devoted to concentrating on solving them. The purpose of exception reporting is to eliminate spending supervisory time reviewing tasks that are on schedule and within budget. Supervisors may then concentrate their time on those tasks that are not conforming to plans.

As an example, let us discuss a project scheduling system with 150 active jobs. Perhaps only twenty of the jobs are not complying to the schedule. This number can be ascertained only by reviewing the schedule for each of the 150 jobs. Review of the data by staff personnel permits the identification of the twenty jobs that are not conforming to schedules. Only these twenty are then reviewed by supervisors. One maintenance department reduced the length of its supervisory review meetings by 55 percent through the use of exception reporting techniques.

The review of the data may also permit the conversion of the information to another method of measurement. For example, labor hour data from the timekeeping system may be converted to equivalent heads.

MAKING DECISIONS

Once data has been reviewed to identify deviations from plans, the control function requires action on the part of supervisors.

Expenditure rates may have to be adjusted to meet budget requirements. Bottlenecks in schedules may have to be alleviated. In some instances the plans themselves may have to be revised. Plans, as well as control practices, should stay in step with constantly varying conditions.

One of the key elements in decision making is to identify the deviation so the supervisors can do something to alleviate the situation. That is, the deviation should be brought to the attention of that level of supervision that can do something about it, rather than to the maintenance manager or a front-line foreman who does not have the authority to make the necessary changes.

The decision-making process is simpler if supervisors who are to execute the decision participate in determining what action is to be taken. There are three reasons for this participation. First, it will encourage commitment on the part of the participants to execute the decision. Second, it will make possible the pooling of the knowledge of the participants in arriving at a viable decision. Third, it will make all participants aware of the rationale behind the decision.

One method of achieving participation is to have a stated time for a meeting on a particular subject involving those supervisors who have the authority to make decisions and execute them. Examples of stated meetings include a meeting every Monday morning of the craft foremen to review construction projects, and a meeting the first week of each month to review budget performance in the preceding month. Another method of achieving participation is to call a special meeting as the need arises, which is often done when it is necessary for supervisors on all shifts to attend.

TECHNIQUES FOR CONTROL

There is a wide variety of methods and units of measure which may be used to facilitate control in a maintenance department. The techniques generally are derived to control the dimensions of time or money or both. Some of the most common are:

- Schedules
- Budgets
- Labor hours
- Labor standards
- Accounting distribution
- Operations integration

APPLYING SCHEDULES

One aspect of control is the use of time. The statement "finish the job by next Monday" is, in essence, an expressed schedule or plan for completion of work within a specified time period. Reliance on schedules as the only control technique is usually not enough. For example, one maintenance manager proudly stated that 98 percent of his jobs were completed on schedule. This statistic sounded good. It meant that the department was meeting most schedules. A deeper examination revealed, however, that this adherence to schedules was being achieved by using large quantities of premium time. Also, parts and materials were frequently requiring special delivery because they were not being ordered in time for normal delivery. The combination of premium-time pay and added delivery costs was causing a marked difference between actual costs and planned cost estimates. Schedules alone were not providing adequate control. Additional techniques that would cover the monetary aspect of control were also required.

USING BUDGETS

The use of budgets as a control technique is a common practice in maintenance departments. Budgetary control involves both cost and time. Costs are expressed in terms of dollars, and the time frame is to be used to control when the expenditures occur. In turn, dollars can be converted to other units of measure techniques, for example, labor hours, headcount, or equivalent heads.

There are various types of budgets that are used in maintenance departments. The three most common are:

- Fixed line item
- Interchangeable
- Variable

Variable Budgets

A variable budget permits the costs expended by a maintenance department to fluctuate in response to some other factor. For example, in a manufacturing plant the amount of direct labor hours may dictate the number of maintenance labor hours that may be expended during a given time period.

The theory behind variable budgeting is that there is a direct correlation between maintenance effort and the operations which this effort supports. Obviously, if this correlation does not exist, it is inappropriate to use a variable budget because the funding for the maintenance effort will not be valid.

The maintenance department in one manufacturing plant has found variable budgeting to be helpful with respect to using premium time. When the manufacturing departments extend their work shifts, the work hours of the maintenance craftsmen in direct support of the operations are also extended on a variable budget basis.

Interchangeable Budgets

Interchangeable budgets permit funds to be reallocated between the various line items within the budget. Control is exercised only upon the total dollar figure of the budget. Thus, an interchangeable budget permits the maintenance manager a certain amount of flexibility in adjusting budgeted expenditures. For example, the maintenance manager could increase the amount spent on in-house labor by using funding designated for purchased labor or materials, as long as the total expenditures remain within the total budget.

Normally there is one restriction within an interchangeable budget—budget dollars classified as expense may not be spent as capital or vice versa. The reason for this restriction is that most organizations in private enterprise and some governmental agencies maintain a separate and more precise set of controls over capital investment. Therefore, capital funding is not interchangeable with expense funding.

Fixed-Line-Item Budgets

Fixed-line-item budgets, as opposed to interchangeable budgets, control expenditures by the amount established for each item or element listed in the budget. Funds are not permitted to be interchanged without a budget revision. Thus, a fixed-line-item budget limits flexibility.

If the budget revision procedure is cumbersome, a fixed-line-item budget generally leads to two occurrences. First, some line items may be overrun during the fiscal period while others are underrun because the world has changed since the original budget was put together and some of the planning assumptions have been negated. When this happens frequently, it will generally lead to the second event which is an attempt to deliberately misclassify expenditures. In order to spend funds for needed effort on line items in the budget where all the money has already been used up, an effort is made to classify the expenditure under line items where funding still exists. This type of creative accounting leads to gamesmanship in the entire budgeting procedure and generally negates much of the purpose of having budgets in the first place.

A good example of this type of circumvention occurred in a manufacturing plant in Arizona. Because of a corporate office austerity program, funding for modification work was severely restricted in the plant. To circumvent this restriction, the plant manager had a lot of modification work done by the in-house maintenance crew under the guise of maintenance work. As a result, valid periodic maintenance and repair work was not accomplished and the buildings and equipment deteriorated. When a new plant manager arrived, he had to get over $500,000 in capital rehabilitation money to restore the plant and equipment and a new periodic maintenance program had to be started.

APPLYING LABOR HOURS

Labor hours are a widely used method of quantifying for measurement. The feedback is usually obtained from time cards which use job numbers to identify the hours expended by each craftsman on specific jobs or projects. The total hours are then compiled for each job by accumulating the hours reported on each time card. The actual accumulated man-hours are then compared to the cost estimate to determine if the job is being completed within budget. More sophisticated methods of man-hour control identify:

1. Hours by craft
2. Equivalent heads by craft
3. Percent of completion in relationship to man-hours expended
4. Premium-time man-hours as well as straight-time man-hours expended

Hours by Craft

The purpose of identifying hours by craft is to determine the specific group that is deviating from plans so that action can be taken to determine why the deviation is occurring in that craft.

Equivalent Heads by Craft

The equivalent-heads-by-craft technique is a derivative of the hours-by-craft technique. In its simplest application, the total hours expended by a craft for a given effort is divided by forty hours to arrive at an equivalent-heads figure. For example, 100 hours equals 2.5 equivalent heads ($100 \div 40$).

A facilities maintenance department that has craft pools for both construction and maintenance work uses the equivalent-heads method for several purposes. First, the level of effort between construction and maintenance work is measured on the basis of equivalent heads.

Figure 17-2 is a sample weekly report. The two major categories listed on the left side of the form are construction and maintenance. This form is used to identify the distribution between these two types of effort.

Week Ending: _____

EQUIVALENT HEADS WORK REPORT

	Electricians	Plumbers and Mechanics	Carpenters	Painters	Total	Percentage
CONSTRUCTION	6.0	7.5	7.2	4.0	24.7	35.2
MAINTENANCE						
Building	3.8	25.1	5.2	6.5	40.6	58.0
Machine Tools	.9	2.3	—0—	.2	3.4	4.8
User Services	.3	.1	.6	.3	1.3	2.0
TOTAL	11.0	35.0	13.0	11.0	70.0	100.0

FIGURE 17-2

SAMPLE EQUIVALENT-HEADS REPORT

Note that maintenance work is subdivided into building maintenance, machine tool maintenance, and user minor services. Requesters are charged for minor services and machine tool maintenance, while building maintenance costs are charged to the maintenance department budget. The second use of the equivalent-heads method is to measure costs against the maintenance department budget as opposed to charges to users.

The department level of effort in these categories listed in Figure 17-2 is reported in equivalent heads for each of the crafts indicated across the top of the report. Deviations between planned and actual effort are readily identifiable. Primarily, this use of equivalent-heads measurement was devised by this maintenance department to assure that one type of activity (e.g., construction) was not causing the level of effort on other activities to suffer. Prior to the inception of the system, construction work was prone to siphon hours from maintenance activities. The report provides indications of such activities so they can be curtailed.

The report shown in Figure 17-2 expresses an equivalent-heads value for each craft by activity and also the percentage of each activity related to the total equivalent heads. Control is achieved when each craft foreman knows what his equivalent-head goal or budget guideline is for each activity. In chapter 20, Figure 20-6 is a sample of an equivalent-heads budget guideline and performance report. The foreman controls the allocation of effort to meet his goal over the entire fiscal period. Therefore, weekly deviations are not considered as important as trends and averaged performance.

Percentage of Completion

Percentage of completion measurement is used to ascertain where a deviation is occurring during the performance of a job.

As an example, let us assume that the cost estimate on a modification project provides 100 man-hours for carpenter labor. If percentage of completion for this craft is reported at 50 percent and fifty hours have been expended, the performance is occurring in conformance with the plan. However, if their work is reported as only being 25 percent complete and fifty man-hours have been expended, then the job is being overrun. Actual costs will exceed the original planned costs.

Measuring Premium Time

Premium time represents a higher expenditure for labor hours than straight time. If the original plan or cost estimate was calculated at strictly straight-time rates, the use of premium time will cause an increase in actual costs over planned costs. This increase would occur even if the total actual hours expended are equal to the planned hours for the job. Therefore, in order to control costs as well as man-hours, a separate measurement of premium-time and straight-time labor hours is utilized.

USING LABOR STANDARDS

Perhaps one of the most controversial issues with respect to managing mainte-nance revolves around the use and types of labor standards for maintenance work. In essence, standards are a measurement method for labor hours. The hours derived from standards are the plan. In turn, the actual hours expended are compared with the planned hours in variance analysis.

In approaching this subject, let us examine the various types of standards that can be used. These are:

- Individual Estimate
- Historical Average
- Work Sampling
- Engineered Standards

Individual Estimate

The term "individual estimate" is generally applied to an estimate prepared by a craftsman or foreman as to how long he thinks it will take to do a job. This type of standard is sometimes criticized because there may be inconsistencies among estimates made by various individuals for tasks that are essentially the same. However, for a small operation, individual estimates may be the only type of estimating that is economically practical. Furthermore, an experienced foreman or craftsman who is knowledgeable of the facilities where the work must be performed may derive a realistic standard for that operation in those specific facilities. If the individual estimate is an accurate form of measurement of performance for that maintenance department in those facilities, it is adequate. There is no need for using another type of labor standard.

Historical Average

Using historical data to develop a standard necessitates some sort of a feedback as to how long it has taken to do a certain type of job. An average time or range of time is developed for the job based upon what has taken place in the past. The main criticism of the use of historical data to develop a job standard is that the approach does not take into account what kind of work methods were used in performing the job. If the methods are poor, the standard is erroneous and the job should be done in less time. For example, if pipe is being manually cut and threaded in the field rather than in the shop using power equipment, the time needed to do the job would obviously be longer, because optimum work methods are not being utilized. However, if the work methods are essentially correct, the use of an historical standard as a method of work measurement is a valid approach.

One facilities maintenance department that operates on historical standards and periodic review of work methods was able to accomplish 84 to 97 percent performance against standards, depending upon the type of repetitive operation to which the standards were applied. This performance facilitated planning, as well as control of operations.

Work Sampling

Work sampling entails the use of the classical industrial engineering time study approach. The work-sampling method of developing labor standards is expensive since time and motion studies require a substantial number of man-hours. The standards derived are valid only if the sampling is representative of how the job is done on a repetitive basis. Further, the standards must be revised if the work methods are changed substantially at some later date. Thus, the primary criticism leveled at work-sampling labor standards for maintenance operations is the cost of the time studies required to establish and maintain the standards.

Engineered Standards

The term "engineered standards" refers to either standards derived from method-time-measurement or other set standards for maintenance work prepared by consulting firms. The general criticism of method-time-measurement (usually leveled by consultants) is that it is too detailed for maintenance operations. The maintenance standards available from consultants are sold on the basis that they are simplified and readily applied. The use of consultant's standards usually involves preparing a detailed description of the sequence of tasks to be performed, as well as assigning the engineered time standards for accomplishment of the job. Some maintenance departments are highly pleased with the use of the engineered standards and methodology provided by a consultant and can demonstrate more than a 20 percent increase in productivity resulting from their use. Others have been less enchanted because of the amount of time required by a planner to prepare a detailed description of the tasks to be performed and to assign the engineered standards to the detailed tasks. The acid test in the use of engineered standards is whether the costs of using them are offset by the savings derived.

The maintenance department for a university decided that it took too many planner labor hours to use the engineered standards supplied by a consultant. As a result, they developed their own standards for certain repetitive operations using historical averages and reduced the amount of labor expended on planning by 20 percent.

USING ACCOUNTING DISTRIBUTION

Another method of controlling costs is provided by the way in which the costs of operating a maintenance department are distributed by the accounting system. The two basic methods of accounting treatment are:

- Service center
- Overhead

Using the Service Center Method

When a maintenance department is set up on a complete service center basis, all costs incurred by the department are redistributed.

The costs of maintaining productive or nonproductive equipment are charged back to the equipment users. Labor is usually charged at an hourly rate for a given craft, rather than at the specific rate paid to an individual craftsman. This hourly craft rate normally includes administrative costs of the maintenance department as well as actual payroll expenses for the craft. Material costs may be charged at actuals or may have a handling surcharge to offset supply room operation costs.

The costs incurred by the maintenance department in maintaining buildings and building equipment are charged back to the departments that occupy the space. This occupancy charge is usually on a rate-per-square-foot basis. Other costs, in addition to expenditures for maintenance labor and materials, may also be included in the occupancy charge rate, such as taxes, insurance, rent, and utility bills.

The idea behind having a maintenance department operate on a service center basis is to make equipment users and space occupiers more cost conscious. Since they are charged for the maintenance work, they are supposed to be more judicious in what they ask for.

The service center method is most effective with respect to productive and nonproductive equipment and modification work. When the maintenance, repair, and rehabilitation of buildings is performed on a service center basis, it may be less effective over the long run. When the space occupier has too much say about the maintenance program, the quality of the maintenance may suffer substantially. Buildings and building equipment may be allowed to deteriorate to the point where eventual refurbishment will be more costly than the cost of regular maintenance.

One major factor to be considered in using the service center method is the amount of data processing involved in billing back each individual job to the requesting organization. With the advent of computerized accounting systems, handling this volume of data is now more readily feasible.

Using an Overhead Operation

As an overhead operation, a maintenance department essentially absorbs all costs against its own budget, and no direct redistribution of the charges is made to those departments receiving the services or occupying the space. The costs of the maintenance department are collected and applied as part of an overhead rate. Equipment users and space occupiers may tend to be less judicious in their demands for service, since the individual work requests are not charged directly back to their departments. The maintenance manager has a direct say in how he will spend his resources since all the money is in his budget. He must establish the priorities in accomplishing work and refuse to perform some work if the expenditures are not covered by his budget.

This method is distinctly different from the service center approach, which essentially places the control of certain expenditures in the hands of the users. Under a service center operation, the requester is responsible for justifying the need for the expenditure and the maintenance department merely provides the requested service.

INTEGRATING OPERATIONS

Another method of achieving control is to assure that the various operations for a given activity are properly integrated. In sophisticated forms, this is sometimes referred to as line-of-balance or critical-path planning. Let's look at a few simple examples.

1. At a college in New England, it was standard practice to hire part-time help in the fall. This help was used to supplement the groundskeeping crews in raking up the leaves that fell each autumn from the many trees on the campus. The crews were to rake up the leaves into large piles, which were then to be loaded into trucks and taken to a mulch pit. However, the rake-up and the truck pick-up operations were not integrated. As a result, the leaves would be raked up one day and wind would scatter them about again on the next. This process might go on three or four times before a truck pick-up would occur. The college had only a limited number of trucks in which to transport the leaves. Additional rental trucks were not readily available. Even if they were, the rentals would have increased the total costs of the activity. Accelerated performance was not required. Time was not a critical factor as long as the leaves were picked up before the heavy snows occurred in the winter. Thus, depending upon when the leaves would start to fall, the operation could be accomplished in around six weeks during September and October of each year.

	WORK SCHEDULE LEAF RAKING IN ZONE # 3 WEDNESDAY AND THURSDAY	
		Activities
Time Periods	**Leaf Raking Crew # 1**	**Truck # 1**
0800	Travel to Work Site	Take Crew # 1 to Site
0815	Rake Leaves	Pit Run
0830		
0845	↓	
0900		↓
0915	Load Truck	Load Truck
0930	↓	↓
0945	Rake Leaves	Pit Run
1000		
1015		
1030	↓	↓
1045	Load Truck	Load Truck
1100	↓	↓
1115	Lunch	Lunch
1130	↓	↓
1145	Rake Leaves	Pit Run
1200		
1215	↓	↓
1230	Load Truck	Load Truck
1300	↓	↓
1315	Rake Leaves	Pit Run
1330		
1345		
1400	↓	↓
1415	Load Truck	Load Truck
1430	↓	↓
1445	Rake Leaves	Pit Run
1500		
1515	↓	
1530		↓
1545	Load Truck	Load Truck
1600	↓	↓
1615	Travel to Yard	Return Crew # 1 to Yard
1630	Clock Out	Clock Out

FIGURE 17-3

SAMPLE INTEGRATION OF OPERATIONS

A simple scheduling routine was introduced, which brought together the trucking and leaf-raking operations on a daily basis by zone. Figure 17-3 is a sample format. This integration of the operations resulted in a reduction in the size of the leaf-raking crews to match the trucking capacity. The repetitive task of raking up the same leaves was virtually eliminated. The total costs of the activity were reduced 65 percent.

2. On a large university campus, the maintenance department accomplished a substantial amount of modification work to meet the requirements of various research projects. As many as sixty percent of the jobs would be started and then had to be stopped because of a lack of materials. The procurement activity was not integrated with the construction activity. The integration was achieved by creating a central planning staff that handled the ordering of materials and their delivery to the job sites. Jobs were not released to construction until the required materials and equipment were available. The number of jobs stopped due to material shortages was reduced 85 percent.

18 Work Reception–Tested Method for Controlling Maintenance Costs

This chapter covers the aspect of control known as work reception. Work reception refers to the way requirements for work are received by maintenance department personnel. Work reception procedures are a means of achieving control over what work is done. The reason for controlling the work to be done is to control the total costs of the operation, which is why this chapter is important. Without control over what work is being planned or performed, a maintenance manager will not have control over costs being incurred. There must be well devised methods for receiving work.

One maintenance manager became an instantaneous success in his new job as head of the maintenance department in a manufacturing plant in Michigan. His success was directly attributable to his interest in work reception. Plant office personnel, including executives, were constantly complaining about the heating and ventilating in the office area. The cause was attributable to defective thermostatic controls. The maintenance foreman responsible was concerned with the expenditure rate of his budget and was only installing one new thermostat control per month. The new maintenance manager decided to buy all of the controls at one time and install them as quickly as possible in order to eliminate the trouble calls and constant complaints. The work was accomplished within thirty days. The complaints ceased and the man received many compliments on the fine job that he was doing as the new maintenance manager. Incidentally, this example is also an object lesson on the matter of conflicting goals. The foreman put the highest priority on his budget expenditure rate. The new maintenance manager considered getting the plant in proper working order as being the more important goal. Since the work would all be done in the same fiscal year and the total costs were essentially the same whether the work was done piecemeal or all at once, the new manager made the proper resolution.

Work reception methods should vary according to the type of operation. Typically, the major operations on which work is performed are:

- Breakdown or service call maintenance
- Periodic maintenance
- Repair work
- Overhaul and rebuilding of equipment
- Construction and rehabilitation of facilities

HANDLING SERVICE CALL MAINTENANCE

Response methods for service call maintenance can be considered to be nothing more than a means of reacting as requests for service are received. This is not necessarily the case. The maintenance manager for one school district believed in planning responses even on breakdown maintenance. He had each of his foremen prepare procedures to cover power failures, water service failure, and sewer stoppages for each school complex. Included in each packet were plot plans and drawings for the water, electrical, and sewer systems of each site. The payback was immediate. One night a site with fourteen buildings had an electrical power failure. Using the electrical drawings, the maintenance crew was able to restore service to thirteen of the fourteen buildings before classes started the following morning.

RUDIMENTARY WORK RECEPTION METHODS

Response methods have varying degrees of sophistication. The more rudimentary methods involve one of the following three levels:

- Individual worker
- Supervisors
- Department staff

Using the Individual Worker

The use of the individual worker is the least sophisticated level at which reaction response is exercised. It can also be the most costly. In essence, the receipt of requests for maintenance service and responses to the requests are left to the discretion of each individual worker. This procedure can result in utter chaos in a maintenance department responsible for a large facility or a multibuilding or multisite operation. Carried to its extreme, there is no visibility over where each worker is, what he is working on, or how to get in touch with him to cover an emergency call. Things run willy-nilly. There is no routing. Man-hours can be wasted in travel time since duplicate trips to a location are made by different craftsmen responding to different requests for service. Calls for vendor service can be made by each worker, which could result in significant expenditures when there is no clear delineation as to how much vendor effort can be authorized by the individual worker. Carried to its ultimate degree, there is no control over the work performed or the costs of getting the work accomplished. Thus, sole reliance on the individual worker to administer reaction response is not warranted.

Maintenance departments that operate with only one or two men on a night or weekend shift may rely on the discretion of the individual worker on sequencing response to trouble calls. However, jobs other than those that can be fixed quickly usually require them to telephone a supervisor for additional instructions.

The need for some control beyond the individual worker's response is not always recognized. One maintenance department devised a highly sophisticated automated system for its multibuilding site. The trouble calls were received centrally and immediately processed into a computer program via a remote terminal. A work order was then immediately generated by the computer program and was printed out at a teletype terminal located in the maintenance office that was responsible for the work in the appropriate geographical area. Thus, there was a rapid method for receiving a request for service, recording the call, generating a work order, and transmitting it to the appropriate maintenance zone office. It was at this point that the system lost control. Each craftsman would go to the teletype and take those work orders that he felt like doing. The rest of the orders remained there until the zone supervisor or shift foreman picked them up and assigned the work. There was no duplicate copy of the work orders at the zone maintenance office so there was no record of what work orders had been received via the terminal until a weekly report was generated by the computer program and distributed to each zone office. The supervisors of the zone office had no immediate overview of all the work that had been assigned to the office or which craftsmen had the work orders. Two craftsmen were frequently working on different work orders in the same building, which resulted in duplicate travel time.

To alleviate the situation, the system was revised. First, two-part paper was installed on the teletype, providing duplicate copies of the work orders, which were retained in the zone office as a record of who was assigned the work. Second, craftsmen were assigned the work orders by the shift foreman, who established the sequence of answering the service requests and a routing to minimize travel time. The use of individual worker response was abandoned.

Using Supervisors

The use of the supervisor as the contact for reaction response does have some advantages over the use of the individual worker. The supervisor can establish the sequence in which responses are made. He can also control which craftsmen are assigned the service requests and can establish a routing to reduce duplicate travel time.

There are disadvantages to using the supervisor. First, the supervisor may not always be available to receive the trouble call. The requester must then wait for the supervisor to return his telephone call or try to seek help elsewhere. The need to seek help elsewhere may cause the requester to feel as if he is getting the runaround. It can also end up by having several different people returning the requester's call, or by having the maintenance manager himself fielding the call. This procedure can waste time both for the requester and the members of maintenance supervision who get involved in the merry-go-round of telephone calls.

A second disadvantage to using the supervisor as the point of contact occurs when the maintenance department has reached the size of being organized by craft or shop. The requester may not know which supervisor or foreman to contact for service. In requesting service he may be shunted from one foreman to another before he gets the right shop. This procedure wastes the requester's time and the time of the maintenance supervisor who must listen to his request, determine which shop should handle the work, and refer the requester to the appropriate foreman.

Using Department Staff

The use of a member of the maintenance department staff as the contact for work requests does have some advantages over using supervisors. At least the staff member knows which shop should respond to the request for service. This system prevents the runaround syndrome. However, as with the supervisor, the staff member may not always be available to receive the call. Again the requester is apt to seek help elsewhere and the merry-go-round of phone calls can begin again.

FORMAL WORK RECEPTION METHODS

Maintenance managers who are responsible for multibuilding or multisite operations cannot expect to rely on rudimentary work reception methods and remain cost effective in their operations. They must move on to more advanced techniques. A large-scale maintenance department that is organized on a craft or shop basis should also use more advanced techniques of work reception. The three methods that provide this higher level of sophistication are:

- Call desk
- Screening and clarification of work
- Feedback

Using a Call Desk

To avoid the disadvantages of having the individual worker, the supervisor, or the staff member as the contact point, many maintenance departments have a call desk. Depending on the nature of the operations supported by the maintenance department, this desk can be manned on a twenty-four-hour basis. Some maintenance departments may operate their call desk on only a one- or a two-shift weekday basis. Trouble calls at other times are made to the security department call desk, which is manned on a twenty-four-hour basis. The security desk contacts the operating engineer on duty or phones a member of maintenance supervision who is on call to have him respond to the request for service.

Another technique to avoid manning the call desk on a twenty-four-hour basis is to use a telephone message recorder when no one is at the maintenance call desk. A recorder permits the requester to leave his message or to find out who to call for service in off-hours.

An experienced call desk receptionist or dispatcher can be a valuable asset to a maintenance department. First, the dispatcher can project a favorable departmental image by handling the calls in a courteous and expeditious manner. Second, if the receptionist is thoroughly knowledgeable of the operations, he or she can often determine the true cause of a problem even though the requester may only be reporting a symptom. For example, the requester may be reporting that fumes are collecting in the degreasing operation. The buildup of fumes is a symptom. The cause is probably an exhaust fan failure.

A call desk receptionist's ability to make an educated guess as to the cause also provides some assurance that the proper person will respond to the trouble call. Also, a capable dispatcher generally knows where all the craftsmen are and can contact the man who is geographically closest to the area requiring service.

The use of radio contact greatly facilitates the dispatch and response time to emergency calls, leads to better control, and increases productivity. One maintenance department increased craftsmen productivity in handling service calls by 15 percent. They made up the cost of the radio system in less than three months.

Screening Calls

The purpose of screening or clarification techniques is to assure that craftsmen with the proper skills and equipment are dispatched to the job site at the proper time. This task entails asking the requester to describe what the situation is or what work needs to be done. A capable call desk receptionist can handle a lot of this effort. In other instances, such as when the travel time is extensive and would involve overtime, or the situation appears highly urgent, the dispatcher may refer the matter to a supervisor for clarification and subsequent authorization to proceed.

One school district was having troubles clarifying work at its various school sites. The maintenance truck would show up to do the work and the men would not be allowed to proceed until the class using the facility had recessed. Lost time per man was averaging four hours per workday. In other instances, the crew on the truck was asked to do all kinds of other work when they showed up to perform on a service order. Performing this additional work meant other work on the daily route was not being done that day. Service order work was constantly performed behind schedule.

The maintenance manager accepted the challenge. First, he established a requirement that a request for work had to specifically state what repairs were to be made and the times the craftsmen would be permitted access to where the work was to be done. Second, he had the principle of each school agree that all requests for work would be made to the resident head custodian. No teachers or school staff were to ask the maintenance truck crews to do any work. The head custodian, in turn, was required to provide a proper description of the extent of the work when he reported it to the maintenance call desk. The results of these improvements in the reception of work became apparent immediately. Within sixty days, lost time was reduced 75 percent and schedule performance increased to better than 85 percent.

Getting Feedback

The purpose of feedback techniques is to determine if additional work should be done. For example, a craftsman may be dispatched on a breakdown maintenance call. He is able to restore the equipment to service on a quick-fix basis. However, additional work, such as replacement of the bearings, may be required. The only way to know that this work is required is to have the craftsman tell someone that it should be done. The feedback provides the notification or reception method for the additional work.

Many maintenance operations still rely on verbal feedback. This procedure is faulty, because verbal reports are frequently forgotten. Written feedback has proven to be the most effective means of assuring that the additional work is properly received and accomplished. One means of getting this written feedback that is frequently used is to have the craftsman note the additional repairs required on the service order he used to perform his work. This procedure eliminates the need for an additional piece of paper for the feedback system.

PERIODIC MAINTENANCE

Periodic maintenance requirements are usually established in one of two ways. The method selected usually depends on the accounting distribution of the costs.

CHARGING THE USER

If the user of the equipment is charged for the periodic maintenance, the user is generally responsible for requesting the implementation of periodic maintenance on the equipment. The user may also review and approve the periodic maintenance work that is to be performed. The request for the work is generally directed to the maintenance manager or craft foreman who supervises the craftsmen who would do the work. A staff planner or the craft foreman then develops the periodic maintenance program and work orders are issued to the craftsmen.

CHARGING THE MAINTENANCE DEPARTMENT

If the items are building and house equipment and the work is charged to the maintenance department, no user of the facilities is involved in establishing the requirements for periodic maintenance work. The requirements are established unilaterally by a staff planner or the foreman assigned maintenance responsibility for the equipment.

HANDLING REPAIR WORK

Generally, requirements for repair work become known as the result of the feedback of information from a service call or a periodic maintenance inspection. Feedback to the responsible maintenance foreman results in a judgment to accomplish the repair.

If the repair effort is extensive, a preliminary cost estimate is generally made before the work is authorized. This estimate should be made to determine if replacement is more logical than repair, particularly with respect to minor equipment items. Some maintenance departments even have standard operating procedures to replace any electric motors rated at one horsepower or less, because it is less costly to buy a new motor than to have a craftsman spending time trying to repair the old one.

If the repair work is usually accomplished in-house, the foreman should make an assessment of the effort that will probably be required prior to commencing the work.

When the work is contracted, it is normally a safe practice to get a preliminary cost estimate from the vendor prior to his accomplishing the repair.

OVERHAULING AND REBUILDING EQUIPMENT

Requirements for the overhauling and rebuilding of equipment become known as the result of reviewing the repair history. The decision to rebuild rather than replace should be based upon the economics of the situation. Since an overhaul or rebuild extends the normal life of equipment, the work is generally classified as being of a capital nature. The foreman responsible for rebuilding and overhauling the equipment should receive the work request.

CONSTRUCTING AND REHABILITATING FACILITIES

For those maintenance departments assigned responsibility for facilities construction and rehabilitation work, highly formalized work reception procedures should be utilized. Without such procedures, the lack of control can result in a lot of wasted time and money because the wrong work is done, the work is done at the wrong time, or materials and equipment are purchased that are not required when the plans and specifications are released. Let's look at some examples.

1. On the basis of a conversation with a design engineer, an air conditioning foreman purchased two water chilling units that were to be used for cooling laboratory equipment. The final design appended a cooling loop to the existing chilled water system within the building and the two units purchased were not required. They sold at a loss of $8000.

2. A foreman in charge of construction purchased $3000 worth of special vinyl wall covering to be used in an area that was to be fitted-up as a marketing office. He purchased the vinyl prior to receipt of plans and specifications, on the basis of a conversation with a member of the marketing staff who was handling the interior decorating. The plans and specifications for the office were never released, because it was decided that the additional space was not required. The $3000 worth of wall covering had already been delivered and now there was no place to use it. Because it was a special order, the vendor was reluctant to accept return of the material. It was finally returned for a 50 percent credit.

3. A project was being planned to fit up a vacant office area in a large manufacturing plant. After a conversation with the project job captain in engineering, the construction foreman decided to go ahead with the demolition portion of the job on a weekend. The crew removed the doors, non-load-bearing walls, and dropped ceiling in the area. Unfortunately, when the drawings were released the following Wednesday, it was discovered that they had torn out four offices that were not part of the modification. Restoration cost over $3000 in materials and labor.

In the above examples, the cause of the goof-ups was a lack of control in the reception of work. The materials and equipment were purchased, or work was done, prior to receipt of plans and specifications or a proper job order. Reception of work for construction and rehabilitation effort is best controlled by having work procedures that require approved plans and specifications and an approved job order for the project prior to any materials being purchased or work being accomplished.

There are instances when, due to procurement lead time, some equipment items may be placed on order prior to release of finalized plans and specifications. In these cases, the authorizing job order should be released and the equipment charged to it. The actual purchase requirement should be documented on an engineering order identifying the long lead time items and stipulating that they will be included in the finalized drawings. The use of the engineering order pinpoints accountability for initiating the procurement action.

19 Modern Maintenance Work Authorization Systems

This chapter covers methods that have been successfully used by maintenance managers in the work authorization systems for their departments.

The term "work authorization" refers to the way expenditures or work are sanctioned within a maintenance department. When controls are too tight, there will be a lack of flexibility that will hamper the ability of the department to be responsive in executing its assigned mission. On the other hand, if there are no controls, unwarranted expenditures are quite apt to occur and the system may be unresponsive because of a lack of coordination regarding what work has been authorized.

There is no one best way of putting together a work authorization system. Each maintenance manager must devise his system and apply those methods which best suit the needs of his particular department.

Basically, there are three levels of refinement in work authorization systems. These are:

- Verbal
- Limited Formal
- Formal

Figure 19-1 depicts these three levels, with the various gradations of refinement within each level. Take a moment to look at it, because Figure 19-1 provides an outline of the contents of this chapter.

VERBAL AUTHORIZATIONS

Verbal methods of authorizing work to be done are at the lowest level of sophistication in a work authorization system. Verbal authorizations usually emanate from three sources: a requester, a maintenance zone or craft shop foreman, or a maintenance staff member. Although the use of verbal authorization may be an expeditious method of authorizing work, it has the following disadvantages:

1. There is no written record of what was done, who did it, and who authorized it.
2. Verbal instructions are prone to misinterpretation of what was requested, which can result in the wrong work being accomplished. Further, a verbal request may even be completely forgotten, which results in the work never being accomplished until a follow-up inquiry is made.

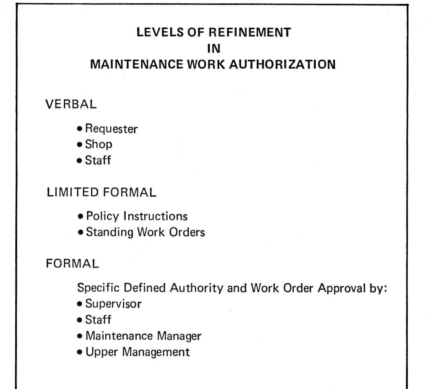

FIGURE 19-1

WORK AUTHORIZATION METHODS

USING REQUESTER AUTHORIZATIONS

The use of verbal authorizations that are not subsequently reduced to writing results in having no history as to the labor hours or material costs resulting from the request. This situation creates a definite lack of knowledge of where and why resources are being expended. It also can create a very cavalier attitude on the part of requesters as to what work they would like to see done. Needs are not sifted from wants, and there is little or no incentive to control costs since they are not identified to a specific individual requester and maybe not even charged to his or her department.

Let's look at some examples.

1. A department store maintenance department made an analysis of the work being performed by their craftsmen on a verbal basis. They were surprised to find that over half their time was being expended carrying stock for the female clerks. This situation made life easy for the clerks, but raised a lot of questions with respect to the efficient utilization of craft labor hours. A work order system was implemented that required the craftsmen to account for their time in performing service calls.

2. A school district made an analysis of the effort being expended by its maintenance crews on a verbal request basis. They discovered that over 25 percent of the time a truck service crew spent at a given school site was devoted to moving things or doing little jobs that were not on the original service request. A practice was established whereby the truck crew chief noted any additional work done on the service order. The maintenance manager then presented the findings to the school superintendent and principals at a staff meeting. The principals agreed to curtail verbal authorizations to service crews and make sure work was authorized on service orders before the crew arrived. This procedure facilitated daily work loading and truck crew routing. The difference between scheduled work and actual work completed was reduced 65 percent.

USING FOREMAN AUTHORIZATIONS

Verbal work authorization by the foreman of a zone or craft shop is a little better from the standpoint of control than a requester's authorization. It is better in the sense that at least a member of maintenance supervision has authorized the effort. However, it still has the shortcomings inherent in any verbal work authorization method. There is no way of seeing what work was done, who authorized it, and who did it.

USING MAINTENANCE STAFF

Verbal authorization from a maintenance staff member is considered to be more refined than one from the requester or foreman, because a maintenance staff member generally acts as a focal point for a specific type of work request, such as office equipment service, lock and key service, or the like. This being the case, all the verbal authorizations are made by one individual for a given type of work. This system provides some level of control and knowledge of what work has been authorized. However, the use of a staff member still does not overcome the shortcomings inherent in any verbal authorization system.

LIMITED FORMAL METHODS OF WORK AUTHORIZATION

These methods constitute the middle level of sophistication in work authorization. The two methods most frequently used are written policy instructions and standing work orders.

USING WRITTEN INSTRUCTIONS

Written instructions provide guidelines as to who may authorize performance of work by the maintenance department or by vendors. The guidelines set forth some general rules to assure that craftsmen and supervisors are not expending resources beyond a certain level without some higher level of authorization. Figure 19-2 is an example of a written instruction for repairs to be performed in response to a breakdown maintenance call. Note that the instruction designates levels of authority delegated to a leadman and the night general foreman.

In smaller maintenance organizations, written instructions may be sufficient, because the maintenance manager only delegates a limited amount of authority. The exposure with respect to expenditures is limited, while the delegated authorization provides flexibility in responding to requests for service.

OPERATING INSTRUCTION

Subject: Authorization of Repair Work on Second Shift

1. Night shift leadmen are authorized to:
   a. Respond to requests for breakdown maintenance service.
   b. Accomplish quick-fix and repairs up to $50.

2. The night general foreman is authorized to:
   a. Call in general vendor service.
   b. Call in vendor service on sewer stoppages in main lines.
   c. Accomplish in-house repairs based upon availability of parts in stock in instances of breakdown maintenance when the costs will not exceed $500.

FIGURE 19-2

SAMPLE WRITTEN POLICY INSTRUCTION

HAVING STANDING WORK ORDERS

Standing work order systems can be applied in different ways. They do have some things in common, however, no matter how they are applied. They provide a written record of the type of work that is to be charged to the work order. They also provide a method for identifying labor hours, materials, or purchased services charged to the standing work order.

Figure 19-3 is an example of the standing work orders used by the maintenance department of a multibuilding university. Note that this application is based upon certain types of equipment. Another way to apply standing work orders is by type of work. In this instance, there would be a standing work order for periodic maintenance, another standing work order for emergency or breakdown maintenance, and perhaps a third blanket order for repair work.

The variations in application of standing work orders involve two methods. The first method is to use only a single work document, the standing work order. The second method consists of using a series of individual documents to authorize work against a blanket work order document.

The use of a single standing work order is obviously the simplest approach. However, this method does not provide a record document of what was done and who did the work in each instance of the performance of work against the standing work order. The total costs of material and labor are merely accumulated as one lump.

The use of individual work orders against the blanket or standing order provides a record of each work authorization. One maintenance department uses the method very effectively on blanket work orders for work charged back to the requesters. The requesting organization authorizes a given amount of funding on an annual standing work order. Individual items of work are then authorized separately at lower levels within the requesting organization, using what is called a minor facilities service request form. Labor and material are charged to the job number assigned to the standing work order. The accumulated minor facilities service request forms can then be reviewed to determine what work was done against the blanket, who did the work, and the labor hours and cost of material used on each job.

The main advantage of standing work orders is flexibility. Additional signature authorization is not required before the response to the request for service. The authorization to respond is already provided by the blanket work order.

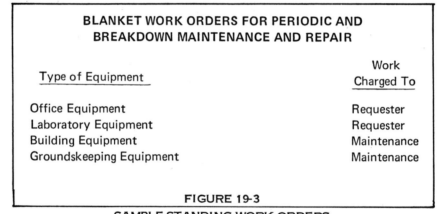

BLANKET WORK ORDERS FOR PERIODIC AND BREAKDOWN MAINTENANCE AND REPAIR

Type of Equipment	Work Charged To
Office Equipment	Requester
Laboratory Equipment	Requester
Building Equipment	Maintenance
Groundskeeping Equipment	Maintenance

FIGURE 19-3
SAMPLE STANDING WORK ORDERS

FORMAL METHODS OF WORK AUTHORIZATION

Work authorization systems are classified as formal methods when there are specific defined signature levels for approving work orders. The levels are usually related to the organizational hierarchy and provide not-to-be-exceeded authorized costs. The lowest level of authorization is the maintenance supervisor or foreman, the next level is the maintenance staff members, the next is the maintenance manager, and the highest is upper management. Figure 19-4 is an example of a signature authorization level listing based upon positions. More refined signature authorization lists cite the names of specific individuals as well as their job titles.

Another method of work authorization is to limit signature authorization by the type of expenditure as well as the cost. Figure 19-4 is also an example of this approach. Note that the not-to-be-exceeded dollar amounts for each job title are different for expense and for capital expenditures. The lowest two levels of maintenance foreman and senior maintenance foreman have a signature authorization for expense, but none for capital expenditures.

SIGNATURE AUTHORIZATION LEVELS FOR MAINTENANCE WORK		
Job Title	Expense	Capital
Maintenance Foreman and Supervisors	$250	None
Senior Maintenance Foreman	$500	None
General Foreman	$5,000	$500
Maintenance Business Manager	$15,000	$1,000
Maintenance Manager	$25,000	$5,000
Division General Manager	$100,000	$20,000
Corporate Office	$100,000 +	$20,000 +

FIGURE 19-4

SAMPLE SIGNATURE AUTHORIZATION LIST BASED ON JOB TITLE

20 Streamlining Maintenance Control Through Performance Analysis

As previously stated, management control consists of assuring that events are conforming to plans. A performance analysis system is a method by which control is facilitated. It is a means to the end of exercising control.

Without a performance analysis system, the chances for a maintenance manager to exercise control are pretty slim. This chapter is important because it defines the essential steps in a performance analysis system and how they should be applied.

DEVELOPING A PERFORMANCE ANALYSIS SYSTEM

An adequate performance analysis system requires the consecutive execution of the following three steps:

1. Comparing accomplishments with plans
2. Examining deviations between accomplishments and plans to determine causes
3. Evolving solutions for deviations

DETERMINING ACCOMPLISHMENTS

To execute this first step, some sort of feedback is required about what has been done. When possible, this feedback should be expressed in the same units of measure as those used in the plan. This procedure makes the data more readily useable for comparison with the plan. For example, schedules are sometimes expressed in terms of weeks rather than specific calendar days. The specific date that a scheduled task is accomplished should also be reported in terms of the schedule week rather than a calendar day. Otherwise, the date will have to be converted to the schedule week before the information can be used in Step 2 of the performance analysis process. In some instances the same unit of measure used in a plan may not be used in the initial feedback. For example, in-house labor information is frequently obtained from the payroll system and is initially reported in terms of hours. These hours must then be converted to man-days or equivalent heads in order to match the unit of measure expressed in a plan.

EXAMINING DEVIATIONS

Step 2 in performance analysis is sometimes referred to as variance analysis because the deviations between accomplishments and plans constitute a variance from the plan. The variance may be positive or negative. Negative variances mean the actuals are greater than the planned numbers. Positive variances mean the actuals are less than the planned numbers. The planned and actual numbers may represent various units of measure such as dollars, hours, equivalent heads, man-days, schedule weeks, or schedule days.

EVOLVING SOLUTIONS

Accomplishing the third step in performance analysis requires a management decision and implementation of the selected course of action. Based upon the causes identified in Step 2, alternative courses of action may be developed and proposed. A management decision must then be made as to which solution to adopt and the selected course of action implemented.

Sometimes the selected course of action may become a standard operating procedure. For example, if the costs on a given job order are going to exceed the original cost estimate, the standard procedure may be to initiate a job order supplement. This job order supplement is then used to revise the original cost estimate and to obtain authorization for the additional funding that will be required to cover the increased costs of the project. Another standard operating procedure would be that when the final actual costs on any given job order exceed the final cost estimate by 10 percent, the causes must be reduced to writing and included in the job order history file. Analysis of this data may then reveal a recurrent symptom or the underlying cause of the overruns.

USING PERFORMANCE ANALYSIS ON BUDGETS

The most common application of performance analysis is accomplished with budgets. There are various levels of refinement in this application that can be classified as follows:

- Informal
- Limited Formal
- Formal

USING INFORMAL METHODS

The informal methods operate on the premise that no news is good news. Someone from the accounting department or upper management must complain that actual expenditures of the maintenance department have exceeded budgeted amounts before the maintenance manager knows that he has lost control of his budget. This approach leaves a good deal to be desired from the standpoint of the maintenance manager executing control.

To start with, the first step (comparing accomplishments with plans) in the performance analysis system is being done outside of the maintenance department. This procedure places the maintenance manager in a defensive position in Step 2 (examining deviations) and Step 3 (evolving solutions). The second drawback is that the examination of the deviations (Step 2) is usually done on a crash basis since the budget is already overrun and pressure is being exerted to resolve the situation before the overrun increases. Finally, the solutions invoked in Step 3 may be highly disruptive to operations because they are also evolved on a crash basis.

A short-term solution frequently applied is a large layoff within the in-house maintenance crew to immediately reduce operating costs, which has a deleterious effect upon morale. It also creates the challenge of having to rebuild the in-house crew to accomplish the department's assigned mission in subsequent years. If headcount must be reduced in a given fiscal period, the need for the reduction should be identified at the beginning of the fiscal period. It is better to lay off one man at the beginning of the year than to have to lay off four or five in the last quarter in order to stay within the total budget by the end of the year.

In a planned reduction program, headcount can often be reduced through attrition and transfer without any layoffs. It is a glaring act of mismanagement when men have to be laid off because a manger has failed to properly exercise control and is ultimately forced to take drastic remedial action to meet an organization's goals. This is often the case when a maintenance manager relies on informal methods of performance analysis.

APPLYING LIMITED FORMAL TECHNIQUES

The two limited formal techniques most commonly used are:

1. Accounting reports of actuals against budget
2. Commitments plus accounting reports of actuals against budget

Using Accounting Reports

A comparison of accounting reports of actual expenditures with budgets shows how much has been booked on the accounting records for a planned expenditure versus how much money has been authorized in a budget. The main drawback in relying strictly on accounting reports is that the information may not be representative of true budget performance because of the lag time that may exist between the time labor and material costs are committed to an effort and the time that the costs are recorded on the accounting records. For example, material may be placed on order, received, and used before the supplier's invoice is paid and the costs are recorded in the accounting records. Depending on how soon the supplier invoices for the material and how soon the actual payment is made, the lag time between the receipt of the material and the recording of the cost in the accounting records may be as long as six months. Thus, the actual expenditures of the maintenance department can be substantially higher than the dollar amounts appearing in the accounting records.

If we measure the time between when an item is placed on order and when the payment is made, the lapse time may be much longer than six months. For example, a new machine in a maintenance shop may have a procurement lead time of 18 months. In such a case, the capital for the item is committed in the year the machine is placed on order, while the actual payment will not occur until the following year, after the machine is delivered.

Using Commitments and Accounting Reports

Because of the lag time in the recording of expenditures in the accounting records, many maintenance departments also use a commitment system, which indicates future costs that will appear in the accounting records. The data indicating what these future costs will be is normally acquired from:

- Contractor bids
- Supplier quotations
- Material pricing catalogs
- Detailed cost estimates derived from plans and specifications
- Budgetary and planning cost estimates

As work is released, the estimated cost is recorded in the commitment system against the appropriate line item in the budget. The summation of all work released constitutes the total committed expenditure against the line item in the budget. Actual costs against committed dollars are then tracked. When the final actual costs have been recorded, the surplus committed dollars are deducted from the total commitment figure. These dollars may now be allocated to other projects. On the other hand, if actual final costs exceed the committed dollars, the additional actual costs are added to the commitment system to assure that a budget overrun will not occur.

Figure 20-1 is a simple format for a manual system used to track commitments and actual costs on job orders. Note that the first column on the left identifies the job by number, the second column is used to post the total committed amount, and the third column is used to record actual costs through the period covered by the report. When all actual costs have been recorded, the fourth column is used to indicate the difference between the committed dollars and actual costs.

Figure 20-2 is a sample of a summary sheet for all job orders. Note that the first column on the left covers the months of the fiscal year. The second column is used to record committed dollars on only incomplete jobs. When the final total costs have been recorded on job orders, the summation is posted in the third column because final actuals should be used whenever this data is available, rather than the estimate or committed dollar figure. The fourth column is used to sum up the committed dollars on incomplete jobs and the final costs on completed jobs. The current total budget figure is entered in the fifth column. The last column entitled "Uncommitted Funds" is used to post the difference between the budget figure in column 5 and the total figure in column 4. If column 5 indicates less money than column 4, the budget will be overrun. If column 5 indicates more money than column 4, the total job orders released to date are within budget.

Using commitments with accounting reports definitely facilitates budgetary control. Here's an example.

Date: _____

JOB ORDER COMMITMENTS
CONSTRUCTION BUDGET

Job Order Number	Total Committed Dollars	Cumulative Recorded Dollars	Final Variance	Remarks

FIGURE 20-1

SAMPLE JOB ORDER COMMITMENT FORM

A new manager was appointed over the maintenance department in a school district. His predecessor used only accounting reports to measure department budget performance, the reports were only generated monthly, and were received thirty to forty-five days after the month-end closing. As a result, there were substantial overruns on some work in the budget, and some unwarranted underruns on other items. The new manager initiated a commitment system, which allowed him to judge the department's performance months ahead of what appeared in the accounting reports and permitted him to exercise some control over his budget. Unwarranted underruns were virtually eliminated as effort was shifted in a timely manner to work in the budget that was running under. Overruns were completely eliminated.

APPLYING FORMALIZED SYSTEMS

In addition to commitments and accounting records of actual costs, formal methods of performance analysis on budgets have two other common factors. First, the responsibility for administering and preparing reports of performance analysis on budgets is assigned to a specific member of the maintenance staff, such as the business manager. Second, the analysis is performed on a continuous basis. If the commitment system is not updated on a continuous basis, the committed dollars on projects could exceed the budget without anyone knowing it until the next update occurred. As a safety measure, many job order systems provide for a sign-off for each job order by the group maintaining the commitment system before the job order can be released.

SUMMARY OF COMMITMENTS
CONSTRUCTION BUDGET

(1) Month	(2) Committed Dollars on Incomplete Jobs	(3) Final Costs on Completed Jobs	(4) Total	(5) Total Budget	(6) Uncommitted Funds
July					
August					
Sept.					
Oct.					
Nov.					
Dec.					
Jan.					
Feb.					
March					
April					
May					
June					

FIGURE 20-2
SAMPLE SUMMARY COMMITMENT FORMAT

Using Projection of Trends

A further refinement in formal systems is the application of projection of trends. The purpose of projection of trends is to identify where deviations of commitments from budgets are apt to occur before they actually happen. This procedure allows more time to develop a solution to the potential deviation and exercise control over the situation.

In some instances the solution may be to defer projects or curtail the effort that would cause the overrun. In other instances, the budget may have to be revised. An example of the latter case occurred when electrical utility costs soared in some parts of the United States due to the increased costs of the oil used in generating the electricity. Few maintenance departments that had budget responsibility for electrical costs had anticipated this significant cost increase. As a result, their budgets had to be revised in recognition of this change.

The use of projection of trends involves the graphing of actual, cumulative commitments and the projection of a trend line. Projection of trends assumes that past events are indicative of future events. Figure 20-3 is a sample projection of trends graph for commitments plotted against an annual budget. Note that the actual, cumulative commitments have been plotted through the month of April using a solid line and have reached $400,000. The budget limit is set at $800,000. The projection is the broken line that indicates that if expenditures continue at the current rate, the budget limit will be reached around August, and by December the cumulative total commitment will exceed $1 million. The projection line indicates that the commitment rate for the year will have to be curtailed sharply if the existing budget limit is to be met.

APPLYING PERFORMANCE ANALYSIS TO SCHEDULES

The purpose of applying performance analysis to schedules may be twofold. First, the purpose may be to control individual job orders that are behind schedule. Second, the purpose may be to determine overall schedule performance on job orders.

CONTROLLING BEHIND-SCHEDULE JOBS

The first purpose is accomplished by analyzing the schedule on each job to determine if the estimated completion date for a given phase has passed and the work has not been completed. The job is then reported as being behind schedule at a certain phase. The report is sent to the supervisor responsible for that phase. The supervisor has three options. First, the work may have been completed by the time he gets the behind-schedule notice, in which case he would post the actual completion date. Second, he can leave the schedule intact and not request a revised schedule. He would normally select this option if he anticipated completing the job in the next few days. Third, he can post a revised estimated completion date and request a schedule change. In this instance the schedule would be revised for other phases of the work as required.

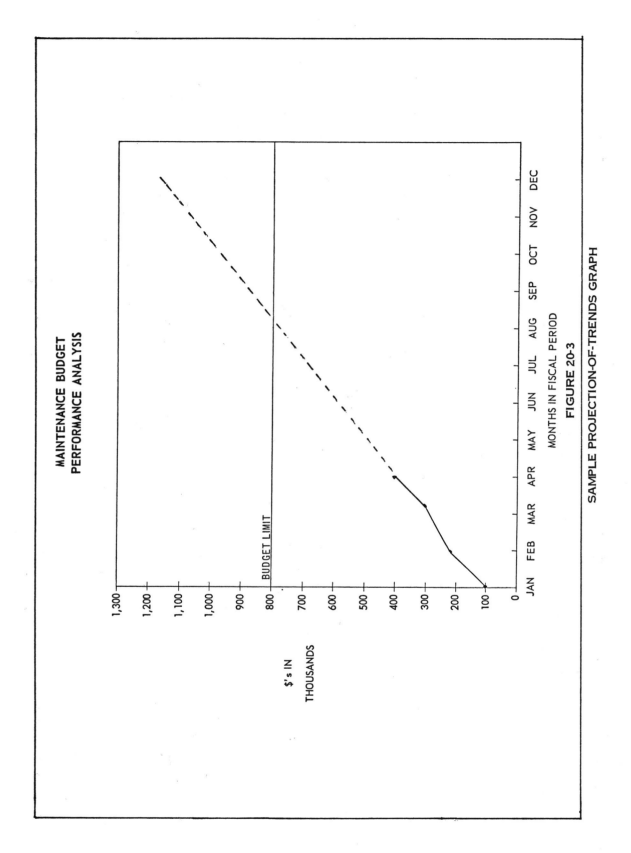

MAINTENANCE BUDGET
PERFORMANCE ANALYSIS

SAMPLE PROJECTION-OF-TRENDS GRAPH

FIGURE 20-3

251

Figure 20-4 is an example of a form used for reporting the behind-schedule status to the supervisor, and for the supervisor's report on what action should be taken. The first column to the left is used to post the job order number. The second column indicates the present estimated completion date. These two columns are filled in before the report is given to the supervisor. The third column is posted by the supervisor to indicate the actual completion date, or the fourth column is used to request a change in the scheduled completion date. The fifth column is used by the supervisor to indicate the cause of the delinquency.

Period Ending: _____

BEHIND-SCHEDULE REPORT
DESIGN

Job Number	Estimated Completion Date	No Action Work Completed On	Supervisor's Report	
			Revise Schedule Completion Date To	Reason for Delinquency

FIGURE 20-4

SAMPLE JOB ORDER SCHEDULE REPORT

DETERMINING OVERALL PERFORMANCE

Determination of overall schedule performance is accomplished by summarizing the performance on all job orders in the system. Figure 20-5 is a sample format of a summary analysis. Note that there are three major categories. The first category is for job orders completed behind schedule. The second category is for job orders completed on revised schedules. The third category is for job orders completed on schedule.

Period Ending: _____ _____

SCHEDULE PERFORMANCE ANALYSIS
DESIGN PHASE

	Completed Behind Schedule				Completed On Revised Schedule	Completed On Original Schedule	Total Jobs
	Type 1	Type 2	Type 3	Total			
Quantity	0	1	1	2	3	15	20
Percentage				10%	15%	75%	100%

FIGURE 20-5
SAMPLE OVERALL SCHEDULE PERFORMANCE ANALYSIS

The first category, covering job orders completed behind schedule, is further refined by subdividing the jobs into three types. Type 1 jobs are jobs completed less than three days behind schedule. Type 2 jobs are jobs completed three to five days behind schedule. Type 3 jobs are jobs completed more than six days behind schedule. The purpose of this subdivision is to indicate the severity of the behind-schedule performance.

The report includes the number of jobs in each category. It also indicates the percentage of jobs in each category in relation to the total jobs completed. The particular system from which this sample format was taken operates on a Friday completion date schedule. Type 1 jobs, jobs completed less than three days behind schedule, are the jobs that were completed that weekend and were released the following Monday to the next phase. Type 2 jobs, jobs completed three to five days behind schedule, are jobs that are finished by the Friday completion date of the following week. Type 3 jobs, jobs completed six days or more behind schedule, are the result of the supervisor failing to request a revised schedule and being unable to complete the work by the Monday of the next schedule week—i.e., the job was more than one week behind schedule by the time he got the work done.

The second category in the format in Figure 20-5 covers jobs completed on time under a revised schedule. This category indicates both positive and negative aspects of the performance. The fact that the schedule was revised is positive since the work was being tracked and schedule relief was instituted when it became apparent that the original schedule would not be met. Further, the revised schedule was met. The negative aspect is that the original schedule was faulty, which is what necessitated the revision of the schedule. The validity of this negative aspect is largely a matter of the type of scheduling system utilized. If the system uses priorities extensively, one could expect a greater amount of schedule revision activity as higher priority jobs are introduced into the system and override schedules on lower priority jobs.

USING PERFORMANCE ANALYSIS ON EQUIVALENT HEADS

The equivalent-heads method measures the level of effort for a given activity. On a weekly basis, the total labor hours charged to a given activity are divided by forty to give the headcount equivalent to these hours. The value forty represents forty hours, which is one man-week or head. For example, 240 hours charged to a given activity by various craftsmen equals six heads (240 ÷ 40 = 6).

The equivalent-heads method is normally used when the same craftsmen must divide their time between various tasks, such as periodic maintenance, repair, and construction work. The purpose of equivalent-heads measurement is to assure that each activity is receiving its proper level of effort in relationship to the budget allocated for that type of work and the other tasks assigned the group.

Figure 20-6 is a simplified example of an equivalent-heads report. The first line cites the equivalent-heads budget guideline for each task. Subsequent lines cite the equivalent heads charged to each task by week, with an average figure at the end of each month. The men in this example are painters who have their total headcount or budget allocated between two types of work. The cost of maintenance painting is charged to the maintenance department. Construction painting is charged to the job orders for the projects. The headcount of the total painting crew is allocated between these two tasks. A deviation from the budget guideline means that labor hours are being underexpended on one task and overexpended on the other. If the budget guideline for maintenance painting is constantly exceeded, the resulting imbalance at the end of the budget period will mean an overrun for the maintenance painting. At the same time, the construction equivalent-heads guideline would be underrun, indicating that the amount of anticipated painting in connection with construction projects was not accomplished.

Feedback on equivalent-heads activity serves to control the level of effort between assigned tasks and to assure that the total crew headcount is at the proper level.

From the data presented in Figure 20-6 it is apparent that the equivalent heads being charged to construction projects has been higher than the budget guideline. The cumulative average at the end of February is 11.7, which is 2.7 higher than the budget guideline of nine equivalent heads. There is a commensurate underrun in the equivalent heads charged to maintenance painting. If this trend remained constant to the end of the quarter covering the months of January, February, and March, the maintenance painting budget would be underrun. The resulting discrepancy would indicate that some of the planned maintenance painting was probably not accomplished. A decision would have to be made regarding what course of action to follow in the next quarter.

EQUIVALENT-HEADS PERFORMANCE
PAINTERS

Budget Guideline	Type of Work	
	11	9
	Maintenance	Construction
Week Ending		
1-3	7.5	12.5
1-10	8.0	12.0
1-17	9.0	11.0
1-24	11.0	9.0
1-31	6.5	13.5
Cum. Average	8.4	11.6
2-7	7.6	12.4
2-14	8.0	12.0
2-21	11.0	9.0
2-28	6.0	14.0
Cum. Average	8.3	11.7
3-7		
3-14		
3-21		
3-28		

FIGURE 20-6

SAMPLE EQUIVALENT-HEADS CONTROL REPORT

USING PERFORMANCE ANALYSIS ON EQUIVALENT MAN-DAYS

In lieu of equivalent heads as the unit of measure, some maintenance departments prefer to use equivalent man-days. Although the unit of measure is different, the technique is essentially the same. To calculate man-days, take the total labor hours charged to a given task during a week and divide by eight. The value eight represents eight hours, which is one man-day. For example, 240 hours equals 30 man-days (240 ÷ 8 = 30).

APPLYING PERFORMANCE ANALYSIS TO LABOR STANDARDS

Performance or variance analysis of labor standards requires the feedback of actual time spent in performance for comparison with the assigned standard time. Causes of variance may then be ascertained. Some maintenance departments require the posting of an explanation of the deviation on the work order either by the craftsman or the foreman. This procedure requires that the standard times be indicated on the work orders. Some maintenance managers do not believe that standard times should be shown on work orders. The general consensus, however, is that they should be indicated. A craftsman has the right to know what performance is expected of him.

Usually, there are two reasons for applying performance analysis to labor standards. These are:

- To validate the standard
- To correct causes of deviations

VALIDATING STANDARDS

Deviations between actual time expended on a job and the standard time assigned to the job may indicate that the standard is not valid. The most usual causes of deviation are changes in the work methods or materials, which result in more or less time being needed to accomplish the job. For example, a maintenance department in Southern California switched to a different type of filter for their building air handling units. This change not only reduced the frequency of the filter changes, but also reduced the time required to change the filters. As a result, the labor hour standard had to be revised to reflect the new required time for a filter change.

There is often a lot of controversy over when a deviation from a standard should be investigated. Obviously, there are bound to be some deviations, because the standard time is based upon how long it should take the average worker or crew to do the entire job under normal conditions following prescribed work methods. All these qualifications leave a lot of room for deviations. As a rule, deviations from most maintenance standards should be justified when they exceed 10 percent. There are exceptions to this rule. For example, a job may be performed normally by two journeymen. If the job is performed by a journeyman and a new apprentice, the actual time may exceed the standard by more than 10 percent because the journeyman is going to spend some of his time teaching the apprentice how to do the job. Also, the apprentice is probably going to spend more time doing his part of the work while he is learning. This situation can result in a substantial deviation from a standard based upon two journeymen doing the job.

One maintenance department does not bother to investigate an individual instance of deviation on any periodic maintenance job that is less than four hours. The normal performance analysis is done by comparing an historical average with the assigned standard. The historical average is computed by an automated system that contains the labor expended each time the operation is performed. The accumulated data are then averaged by the computer program and reported with the assigned standard for comparison.

The idea behind this approach is that the standard is an average time; therefore, the actual time should also be averaged. For example, the lubrication of a machine tool may have a standard of five-tenths of an hour. The actual performance one time may take three-tenths of an hour because no oil has to be added. The next time the job may take seven-tenths of an hour because lubricants have to be added. In both instances, the deviation amounted to 20 percent. However, the average of the two performances equals the five-tenths of an hour.

CORRECTING CAUSES OF DEVIATIONS

The two most common methods of correcting deviations from standards are to train or retrain personnel in proper work methods or to change the work methods.

Obviously, if a craftsman forgets to bring along the proper tools and materials, or does not follow the proper instructions for assembly or disassembly, a job is going to take him longer to perform. It could also result in damage to an equipment item. Therefore, the execution of management control leads to training.

Changes in work methods may also be required, such as increasing the quantity of a required tool crib item in order to facilitate availability. In one maintenance department, the performance on the servicing of lights in one zone was repeatedly reported as one-half to three-quarters of an hour longer than standard. Investigation revealed that the additional time was expended in moving wooden ladders from one building to another. The solution was to change the work method by buying additional ladders so that one was available in every building.

21 Successful Applications of Automated Information Systems in Maintenance

This chapter presents an overview of some of the applications of automated information systems for maintenance departments. It tells how these systems have been applied to reduce operating costs and to facilitate management control.

Whether information is compiled on an automated or manual basis, one of the major purposes of the compilation of information is to support management control, which is why the subject of automated information systems appears in this section pertaining to control.

Automated information systems should be more than something nice to have. They should be a viable part of the program for the execution of control by providing timely and meaningful information about what is going on.

Information on activities should not be restricted to present daily operations; it can also be applied to what the planning looks like for future actions. Automated information systems can facilitate control of planning and staffing, as well as provide information on daily operations.

The use of automated information systems to support performance of maintenance operations is increasing. In many instances the effectiveness (cost-wise and performance-wise) of maintenance activities can be significantly assisted through the judicious use of computer technology. A maintenance department at an engine test site in Southern California reduced annual operating costs by over $38,000 through the use of an automated scheduling and forecasting system for periodic maintenance. These savings were predominately in labor and were largely achieved by the automated grouping of work orders by zone to reduce travel time and the judicious adjustment of the time intervals applied to the various periodic maintenance operations.

USING A COMPUTER

Properly applied, an electronic digital computer is a useful tool for managing maintenance operations for the following three purposes:

1. Manipulation of data
2. Rapid information transmittal and retrieval
3. Standardization of reported data

MANIPULATING DATA

An automated system can be a powerful tool because of its capability to store and manipulate a vast amount of data. Its use can result in a direct savings in clerical labor costs. One maintenance department was able to eliminate eight clerical positions as the result of instituting an automated work order system covering periodic maintenance operations. This reduction resulted in an annual savings in labor and related payroll expenses of over $55,000.

TRANSMITTING AND RETRIEVING INFORMATION

On-line automated systems can provide virtually instantaneous (real time) transmittal and retrieval of information via remote terminal inquiry. Even the more traditional batch control information systems can provide written reports in a wide variety of formats on an overnight basis.

A manufacturing plant in Michigan makes inquiries into its maintenance work order data base to identify machine tools that should be considered for rehabilitation or replacement due to the total costs for maintenance and repair. The report formats may be changed through the use of a generalized report-writing inquiry program with overnight service.

STANDARDIZING DATA

Computerization provides standardization of reported information, which is particularly significant for a multisite maintenance operation where manual reports are prepared by different clerical personnel at various locations. Manual reports from the different locations can often vary in format, information included, and the period of time covered. These differences can make the comparative analysis of the data difficult and the drawing of conclusions from the data dangerous.

An automated system used by one company covers maintenance operations in six different locations in the United States. All the data is stored in a single computerized data base for reporting in standard formats.

CATEGORIZING SYSTEMS

Some systems analysts have found it convenient to segregate the application of automated information systems between (1) the doing or operating activities and (2) the planning or pre-doing activities.

PERFORMANCE SYSTEMS

In the performance or operating area of maintenance, we are concerned with information regarding the carrying out of activities. This area would include systems for job orders on facilities construction or for maintenance and repair activities. The information in these systems may pertain to budgets and costs for activities (including payroll and material expenses), schedules, inventories, and documents authorizing work.

PLANNING SYSTEMS

In the planning and problem-solving area, activities are different from those of the operating area. The essence of what goes on in this area is exemplified by activities such as cost forecasting and manpower planning, which require searching for alternate directions and selecting courses of action to pursue from among identified alternatives.

COMBINED SYSTEMS

Although it may be convenient to initially categorize the application of an automated information system into either the operating or planning area, one should not overlook the significant interrelationship between the two areas. The data collected in performance can have significant impact on planning. The computer systems supporting operations have the means to collect and analyze data fast enough for the information to be useful in planning. For example, if the expenditure rate of premium-time man-hours reported by a system in the operations area is substantially above the planned rate, early knowledge of the deviation would provide more time in which to seek a solution and make required adjustments to plans. One maintenance department receives weekly reports from the payroll system which identify premium pay hours by craft for comparison with plans. The reported data also identifies the man-hours expended for various efforts such as periodic maintenance for each craft. Reporting of data on a weekly basis permits immediate required action to be taken in the case of deviations.

A maintenance work order system used by one maintenance department combines both doing and planning capabilities. The system automatically generates work orders at the prescribed times for periodic maintenance, which supports operations. The system can also forecast what periodic maintenance work orders will be generated for each craft, which supports the planning. The standard labor hours data in these forecast reports is used in the preparation of annual budgets, manpower planning, and level of effort planning.

APPLYING SYSTEMS

There are a variety of computerized information systems being used to support maintenance operations. To give a general idea of potential applications, this section provides some brief descriptions of five different types that have proven to be effective. These are:

- Job Order Systems
- Maintenance Work Order Systems
- Cost Accounting Systems
- Standard Stock Inventory Systems
- Building Automation Systems

JOB ORDER SYSTEMS

The application of the computer in the area of maintaining an historical data base of effort performed on job orders is being effected by larger organizations that have their own computer. The advanced systems trace job orders through their various steps to completion, compare scheduled to actual days of performance, and calculate the variance between planned time and actual time at each step. Others also include data on actual expenditures versus cost estimates for labor, material, and contracted services. These systems handle job orders that represent substantial expenditures involving funds of a capital nature. Some of the systems are integrated with the accounting system and have the additional ability to prorate costs to expense and capital account numbers.

These job order systems are used in executing control of schedules by using performance analysis. They are also used in budgetary control and planning commitments. Using such a system, a maintenance department in one manufacturing plant reduced its behind-schedule status by 20 percent. It also was able to rapidly identify projects being overrun and underrun. Unexpended funds on completed projects were immediately reallocated and committed to other projects during the year. This procedure not only assures performance within budget, but usually permits the reallocation of some $75,000 to $150,000 worth of funding annually to additional needed projects that had not been included in the original budget.

Another large manufacturing plant uses a job order system to control over 3500 jobs orders annually representing an expenditure in excess of eight million dollars.

MAINTENANCE WORK ORDER SYSTEMS

These systems provide an historical data base for maintenance and operations work performed. They contain data on a large number of work orders for maintenance and minor repair activities of an expense nature. Although each maintenance work order by itself represents a small expenditure, the total value of the work orders in the system may be substantial.

The automation of production leads to more exacting maintenance standards since a breakdown will affect the whole line. In turn, the application of these more exacting maintenance standards can be facilitated by computerized scheduling of periodic maintenance with the automatic generation of maintenance work orders at the prescribed times. Some of these maintenance work order systems have been developed to include the generation of work orders to perform periodic work.

Work order systems may also report on periodic work that has not been performed within the prescribed time period. These backlog or delinquency reports can be used to measure effectiveness in accomplishing the work and to identify areas requiring improvement.

These systems may also provide man-hour forecasting for each craft skill through the application of a labor standard to each periodic operation. This forecasting capability of the system may be used in the nonoperational (planning) area, as well as by the frontline foreman in day-to-day operations.

In a periodic maintenance program of substantial scope, a manual system rapidly reaches the point of becoming unwieldy. One automated maintenance work order system contains over 14,000 separately scheduled items of work and generates over 40,000 work orders annually. The manual preparation of the work orders for this amount of periodic maintenance would require a large clerical staff. A manual follow-up system to assure that the work was performed and the manual posting of work history to an individual item record would also entail extensive clerical effort. The use of the computer eliminates the major portion of this manual effort and reduces annual clerical labor costs by over $80,000.

COST ACCOUNTING SYSTEMS

The automation of cost accounting information for maintenance departments has usually been set up as a part of the automation of the parent organization's accounting system. As a result, these systems are sometimes completely oriented to the needs of the organization's accounting department. The chart of accounts used does not meet the needs of the maintenance department. Often many different cost elements are lumped into one account number and the detail is lost. The separate maintenance department activities for which individual cost control is desired are no longer discernible. The data is not available for the purposes of identifying causes of the expenditures or for planning for such expenditures by separate activity in the future. There are several solutions to this predicament.

One solution is to work with the accounting department in revising the accounting system to provide the required availability of data. Depending upon the automated system, this revision may be readily achieved by adding subaccounts to the existing account numbers or revising the chart of accounts to add additional account numbers, which can be pyramided for organization cost accounting purposes.

Another solution is to have representatives of the maintenance department actively participate in any new accounting system that is being developed. The maintenance departments in one state university system were able to get the cost data that they needed by adapting this approach.

A third solution is to include coding in the maintenance work order or job order system which provides the detailed cost data that is not available in the organization's cost accounting system. Figures 21-1 and 21-2 are examples of craft codes and identification numbers used in one maintenance work order system to provide information on the type of work done and which craft did the work. These codes are applied to each maintenance work order. The system reports the cost of work in several ways. One report sorts first by craft and then by work identification number to provide information on craft activities. The other report sorts first by work identification number and then by craft to identify total costs for each work identification number. Costs are identified as being either material or labor.

Figure 21-3 is an example of work identification codes used in a job order system to provide detailed cost information by the type of work done on job orders.

```
┌─────────────────────────────────────────────────────────────┐
│                        CRAFT CODES                          │
│                 MAINTENANCE WORK ORDERS                     │
│                                                             │
│   Code                    Description                       │
│   001            Machine Tool Machinist                     │
│   002            Machine Tool Oiler                         │
│   100            Electrician                                │
│   201            Air Conditioning Mechanics (1st Shift)     │
│   202            Air Conditioning Mechanics (2nd Shift)     │
│   203            Air Conditioning Mechanics (3rd Shift)     │
│   204            Sheet Metal Men                            │
│   300            Carpenters                                 │
│   400            Painters (1st Shift)                       │
│   401            Painters (2nd Shift)                       │
│   500            Plumbers                                   │
│   600            Gardeners                                  │
│   700            Janitors                                   │
│   800            Office Equipment Repairmen                 │
│   900            Support Services Personnel (supply, tool crib) │
│                                                             │
│                                                             │
│                      FIGURE 21-1                            │
└─────────────────────────────────────────────────────────────┘
```

SAMPLE CRAFT CODES

STANDARD STOCK INVENTORY SYSTEMS

These systems are utilized for inventory management of standard stock items used repeatedly by a maintenance department, such as electrical, plumbing, carpentry, janitorial, and painting supplies. The data elements consist of a stock number, description, quantity on hand, quantity on order, quantity issued, unit price, and minimum stock balance. Quantities ordered, received, and issued are posted for each stock number. The automated system reports when a reorder should be made, based upon the prescribed minimum quantity. When the system also contains the date of last issue, it is possible to identify inactive line items that may be reviewed for purging from the standard stock inventory. The system may also be used to price out individual requisitions by automatically extending the quantity issued times the unit price and totaling all line items on each requisition.

One maintenance department has such a system in an on-line, time-shared mode. This system permits update through a remote terminal and the immediate reporting of inventory shortages for action. With this timely reporting, inventory shortages were reduced by 80 percent during the first six months of operation.

In determining how much should be spent developing an inventory system, some managers use a 3 percent rule. First, they determine the approximate value of the inventory. Next, they determine 3 percent of this valuation. The resulting figure is what should be spent to develop the system. For example, assume the inventory is valued at $200,000. Three percent of $200,000 is $6,000. The development cost should not exceed $6,000. The rationale behind this rule is that no more than a 3

WORK IDENTIFICATION CODES
MAINTENANCE WORK ORDERS

Site

10	Grounds
11	Parking Lots and Driveways
12	Lawn Sprinkler System
13	Parking Lot Lights
14	Painting—Road Stripes, Curbs and Stop Signs

Buildings—Structural

20	Roofs
21	Painting—Interior
22	Painting—Exterior
23	Floors
24	Walls
25	Windows and Blinds
26	Doors
27	Ceilings

Buildings—Electrical

30	Lighting System
31	Power System
32	Telephone, Alarms, and Communication Systems

Buildings—Mechanical

40	Refrigeration System
41	Heating System
42	Air-handling System
43	Plumbing System—Including Sewers and Fire Sprinklers

Building Services

50	Janitorial Services
51	Fire and Safety Devices
52	Material and Equipment Handling

Supply Operations

60	Storeroom Operations
61	Tool Crib Operations

FIGURE 21-2

SAMPLE WORK CODES

JOB ORDER
WORK IDENTIFICATION CODES

01	Facilities Construction and Rearrangements
02	User Machinery and Equipment Installations
03	User Machinery and Equipment Repair or Rehabilitation
04	Miscellaneous Personal Property Fabrication/Repair
05	Building Equipment Installation
06	Building Equipment Repair or Rehabilitation
07	Site Improvements
08	Site Maintenance and Repairs

FIGURE 21-3

SAMPLE JOB ORDER WORK CODES

percent expenditure of the inventory value is warranted to achieve automated inventory control.

Another useful method is to take 1 percent of the dollar value of inventory turnover as the limit. For example, let us assume that the annual dollar value of all standard stock items issued is $800,000. One percent of $800,000 equals $8,000. It would be appropriate to spend no more than $8,000 in developing an automated system.

BUILDING FACILITIES AUTOMATION SYSTEMS

Computer-based building facilities automation systems for control of lighting, heating, air conditioning, ventilating, energy management, security, and protection of buildings can have a marked impact on maintenance management. One of the most tangible sources of operating savings from building facilities automation can be in the reduction of maintenance costs. In some instances, the cost of a complete facilities automation system can be economically justified on the basis of savings in maintenance costs alone. Prior to the escalation of energy costs and limitations on the availability of power, reduction in maintenance costs was the most salient reason for installing these automation systems.

A computer-based automation system for building equipment can be used to:

1. Detect failures
2. Log equipment run-time
3. Automatically schedule periodic maintenance based upon run-time
4. Accumulate data on equipment malfunctions and repair frequencies
5. Reduce maintenance inspection tours

Detecting Failures

Usually, an on-off status device with a signal alarm for the occurrence of a failure is used for failure detection. More sophisticated systems endeavor to provide a warning when a piece of equipment is beginning to fail, rather than wait for an actual failure to occur. An example of incipient failure detection would be to measure temperature or vibration for a major air-handling unit to predict a failure due to worn ball bearings for the drive shaft.

Logging Equipment Run-Time

Actual running time of equipment can be logged by the automated system. This is sometimes referred to as totalization. A device is attached to each piece of equipment for which totalization is desired. When the equipment is running, the automated system clocks the time in seconds, minutes, and hours. This data is stored on a daily basis to compile a complete usage history. The compressor for an air conditioning system is an example.

Scheduling Periodic Maintenance

Based upon run-time, the automation system can generate a report or an actual work order for a given piece of equipment for performance of periodic maintenance. For example, an oil change for an air conditioning compressor is often based on the number of hours of operation. When the compressor has actually operated that number of hours, the automation system would report that an oil change is necessary. When actual run-time data is not available, the normal practice is to use arbitrary calendar plans. In some cases, the use of actual run-times as the basis for periodic maintenance can reduce the work load. For one maintenance department, the switch from a calendar plan to actual run-times reduced man-hours for periodic maintenance by 7 percent.

Accumulating Data on Equipment Malfunctions and Repair Frequencies

The accumulation of data on equipment failures and repair frequencies can serve three purposes.

First, it can permit comparative analysis of similar equipment from different vendors, which can enhance the selection procedure in the selection of equipment in the future. Second, it can identify equipment that is causing the most trouble. Equipment troubles can sometimes be remedied through equipment modifications. In some instances, unscheduled downtime has been reduced 25 percent by using this approach. Third, the data may be used to increase the adequacy of the spare parts inventory for the equipment.

Reducing Maintenance Inspection Tours

When the central automation system is applied to a multibuilding site, savings can be achieved through a reduction in the number of maintenance tours. Since the equipment is being monitored by sensing devices, the need for visual checks is eliminated. One maintenance department was able to reduce its weekend crew coverage by 50 percent. A school district achieved an annual savings of $24,000 by a reduction in tours.

This cost reduction is not limited to strictly multibuilding sites administered by one maintenance department. One company provides a building automation service that covers twenty-five separate buildings in the greater Los Angeles area, utilizing a single computer installation with leased telephone wires connecting it to the various buildings. Smaller maintenance departments may reduce their coverage or inspection tours where there is participation in a shared central system. In addition, all users of the central building automation service have reduced their energy costs by more than 25 percent.

FACILITATING PERFORMANCE ANALYSIS

As previously discussed in chapter 20, the first step in a performance analysis system is the comparison of accomplishments with plans. The second step is the identification of deviations. Both the comparison and the identification of deviations can be greatly facilitated by an automated information system. The system may receive and store the data and automatically report deviations for analysis. The automation provides discipline in reporting data and speed in compiling the feedback and identifying deviations.

In one maintenance work order system, the actual labor hours used in performance of work are fed in. These labor hour statistics are first used in redistributing the costs of the work to requesters. They are then retained for use in performance analysis. A report provides an historical average for each periodic operation for comparison with the assigned labor hour standard. This report permits an evaluation of the assigned labor hour standard for the periodic operations without the necessity of expending a large amount of clerical labor hours in compiling an historical average labor hour figure for each periodic operation.

PLANNING APPLICATIONS

As previously mentioned, an automated maintenance work order system can be used as an effective control tool in planning when the system has a forecasting capability.

Forecasting reports can be designed to support many planning activities. For example, one maintenance department has a forecast report categorized by the craft skill required to perform the scheduled work. Within each craft skill the report indicates the periodic work by each week in which the work orders will be generated by the computer program. This report is useful in determining man-hours required by each craft skill to perform work in the weeks covered by the report. The report is also used to determine if the workload of periodic maintenance to be performed by the craft skill has been properly distributed and pinpoints the need for adjusting the workload schedule.

Another report can sort the scheduled work by building number for each craft skill. The report is used to determine if the work has been scheduled properly by location to minimize travel time, and to pinpoint the need for adjusting schedules to support this end. This technique has saved one maintenance department over $80,000. This same report is also used to determine the amount of periodic work that will be eliminated if a given building or station is dropped from the facilities inventory. This procedure has enhanced facilities planning. It saved $40,000 in maintenance costs one year.

Forecasting reports should be flexible relative to the period of time covered in the report. The reports should be able to cover any period in the future. This goal can be accomplished by an indication of the start date and end date for the report. This flexibility permits the generation of a forecast report based upon its purpose. For example, the maintenance supervisor may desire to know all work operations scheduled in the next quarter in order to plan vacations or determine man-hours by craft skill required for scheduled work. Thus, the requested time period would be for the three months in the quarter.

Another purpose of the forecast report may be the preparation of an annual budget. In this case, the time period of the forecast would be for the twelve months included in the fiscal year.

In another instance, the report may be for a ten-year period. One maintenance department uses this extensive forecast period to plan the complete painting cycle on its multibuilding site.

STAFFING APPLICATIONS

When the personnel records are automated, they can facilitate performance of the management function of staffing. Some maintenance departments use their automated records to get age profiles by craft for the purposes of replacement planning and apprentice hiring. Another use of automated records is the development of ratios in each craft of journeymen to apprentices.

A more conventional use of automated personnel records is to get actual strength reports. There are always a certain number of craftsmen on medical leave who will not be available in planning day-to-day assignments. Another more conventional use is to perform vacation planning using accruals in the automated records.

22 Avoiding Pitfalls in the Automated Maintenance Information System

Automated information systems are sometimes subjected to bitter criticism, usually because of operational difficulties encountered by the users of the system. Much of this criticism frequently stems from a lack of understanding by the users as to how to operate the system. Normally, there are two reasons for this lack of understanding. The first is a failure to have adequate user participation in planning and testing the system. The second is a lack of an adequate implementation program.

The people in a maintenance department who will use the system must have a clear understanding of both the limitations and the advantages of a new automated information system. Then there will be no misconceptions about its capabilities. A clear understanding is achieved by participation in a step-by-step procedure that should be followed in designing and operating an automated information system. This chapter is important because it describes that procedure.

If the steps described in this chapter are not followed, you will probably end up with a lot of frustrated people and an automated system that will not be all that it should be. Further, the costs of development and implementation will probably be 50 to 100 percent higher than they should have been. Do it right the first time. There is a distinct difference between subsequent enhancements to a good system and having to completely rework a bad system in order to get what was originally desired.

As discussed in the previous chapter, automated information systems in maintenance and operations may be widely applied. Regardless of the application, the steps described in this chapter are necessary for developing a successful system. Listed in chronological order, these six steps are:

1. Operations Analysis
2. Requirements Analysis
3. Design Concept and Specifications
4. Program Development and Test
5. Implementation
6. Evaluation

TIMING THE PHASES

The chronological sequence indicated above does not mean that activities under one step may not begin prior to completion of the previous phase. Actually, an overlapping of the phases is normally effected in an effort to shorten the total time required to develop a system. Too much overlap, however, can be counterproductive because, in essence, each phase provides the information necessary for accomplishing the next step. A particularly critical area is the commencement of actual programming in the program development and test phase prior to the completion of the design concept and specifications phase. Unless the design and specifications are fairly well finalized, actual programming should not begin. If any programming is done, it may prove to be a premature effort and considerable reprogramming may be required to meet the finalized design and specifications. Reprogramming can cause a substantial increase in the costs of developing the system because of the added labor costs.

USING SYSTEMS CRITERIA

There is no perfect formula for developing a completely successful automated information system. However, the application of the following tests of quality during the six phases listed above will definitely increase the chances of success.

1. Cost effectiveness—achieving cost savings that exceed the cost of developing and operating the system
2. Veracity—assuring that the information in the system and that reported from it is accurate
3. Viability—having the ability to handle changes and anticipated growth
4. Relativity—meeting the actual needs of the using organizations
5. Timeliness—having a response time that meets the needs of the using organization

Each of the above criteria is applicable in all phases of developing the system. Certain standards, however, should be more heavily emphasized during specific phases to help assure the success of the system. Figure 22-1 indicates which criteria should be emphasized in each of the six phases of development. Specific activities that should be performed to support these five criteria are described in the following sections.

STEP 1—THE OPERATIONS ANALYSIS PHASE

Irrespective of the proposed application, an operations analysis should be performed first. This first phase is a profile check, which consists of determining how things are currently done and what information is currently being processed. The primary purpose of the operations analysis phase is to determine if there really is a need for an automated information system and if such a system would be cost effective.

PERFORMING THE PROFILE CHECK

Figure 22-2 is an outline that may be used to plan the scope of an operations analysis and to classify information after conducting the analysis. This outline begins by first identifying the five major management functions: planning, organizing, staffing, controlling, and directing. Within each of these functions, certain operations are

**RELATIONSHIP OF CRITERIA AND DEVELOPMENT PHASES
FOR AUTOMATED INFORMATION SYSTEMS**

Phase in Development	Standard to be Emphasized
Operations Analysis	Cost Effectiveness
Requirements Analysis	Relativity and Cost Effectiveness
Conceptual Design and Specifications	Viability, Timeliness, Veracity
Programming and Testing	Relativity, Veracity, Viability
Implementation	Veracity
Evaluation	Cost Effectiveness, Relativity, Veracity, Viability, and Timeliness

FIGURE 22-1

APPLYING QUALITY CRITERIA DURING SYSTEM PHASES

identified which pertain to the execution of the function. For example, under the planning function there are the six operations of budgeting, level of maintenance, identification of work, job planning and estimating, scheduling, and maintenance contracts. Each of the operations should be examined to determine:

- How is it currently being performed?
- Who is performing it?
- When is it being performed?
- What is the current level of sophistication, and should it be improved?
- Where would automation cost effectively facilitate the operation?

As an example, let us assume that the operations analysis included the scheduling operation listed under "Planning" in Figure 22-2. The analysis reveals that there is a scheduling system for both major jobs and periodic maintenance. The scheduling and preparation of the work orders and reporting pertaining to the system are all accomplished on a manual basis. The analysis further reveals that the number of active major jobs averages 100 at a time. One clerk on the first shift spends an average of fifteen hours per week to schedule and prepare status reports on these major jobs. On the other hand, the periodic maintenance scheduling involves 4000 machine tools with an average of three scheduled operations per machine and 2000 items of building equipment with an average of two scheduled operations per item. The preparation of the work orders at the appropriate scheduled frequencies requires eight full-time clerks on the first shift. There is no reason to automate this scheduling system for major jobs, since there would be no substantive reduction in clerical labor. For the scheduling of periodic maintenance, however, the automation would be cost effective, because part of the work of the eight clerks could be performed less expensively by an automated system.

MAINTENANCE MANAGEMENT OUTLINE

MANAGEMENT FUNCTIONS	ACTIVITY
Planning	Budgeting
	Level of Maintenance Effort
	Identification of Work
	Job Planning and Estimating
	Scheduling
	Maintenance Contracts
Organizing	Organization Structure
	Assignment of Responsibilities
Staffing	Manpower Requirement
	Manpower Assignment
Controlling	Work Reception
	Work Authorization
	Performance Analysis
Directing	Coordinating
	Communicating

FIGURE 22-2

SAMPLE PROFILE CHECK

PREPARING COST FIGURES

Upon completion of an operations analysis, realistic cost estimates for subsequent systems development and operation should be prepared together with anticipated tangible cost savings. If the tangible savings appear to be greater than the anticipated costs of the automated system, the development should proceed to the requirements analysis phase. If there is no evidence of cost effectiveness, the project should be abandoned to avoid incurring the additional costs of conducting further phases. To assure that all costs of the automated system have been considered, the following major cost elements should be quantified.

Systems Development Cost Elements

- Labor required to perform the requirements analysis, design concept and specifications, program development and test, implementation, and evaluation phases
- Computer and peripheral equipment time for systems test
- Printing of source documents for the test phase

Operational Cost Elements

- Source document input labor cost (clerical, keypunching, remote terminal operation)
- Computer and peripheral equipment time, including any remote terminal installations
- Labor cost for programming and evaluation support to the system
- Printing of source documents for full-scale operation

STEP 2—THE SYSTEMS REQUIREMENTS ANALYSIS PHASE

After the operations analysis has been completed and a decision to continue the development has been made, the next step is to conduct a systems requirements analysis. The primary purpose of this phase is to determine what functions the automated system is to accomplish and how it will facilitate the operations performed by the maintenance organization. An adequate requirements analysis should accomplish the following:

- The identification of the data elements that are to be automated
- A determination of how the data elements are to be automated
- A determination of what is to be done with the data after input—i.e., what type of reports are required and how long the data should be retained

PREPARING THE SYSTEMS REQUIREMENTS ANALYSIS REPORT

Figures 22-3A and 22-3B constitute examples of a systems requirements analysis report for a construction labor hours system. They present a simplified version of a system utilized by the maintenance department of a large university located in Southern California. Note that the report is divided into four sections covering the objectives of the system, a summary of the functions of the system, a description of the steps leading up to the automated processing, and a matrix indicating the data elements, input (source) documents, and output reports required by the system. The report covers the three activities that should be accomplished in a requirements analysis phase to assure that the proposed system meets the criterion of relativity.

REVISING COST FIGURES

Upon completion of the requirements analysis phase, the criterion of cost effectiveness should be applied again, prior to proceeding to the next step. Based upon the more definitive information acquired in the requirements analysis phase, the original budgetary and planning cost figures prepared at the end of the operations analysis phase should be refined together with the anticipated cost savings figures.

AUTOMATED CONSTRUCTION WORK ORDER CRAFT LABOR SYSTEM

Objectives

1. To capture and maintain estimated and actual labor hours by craft for a given work order and to generate reports.

Summary of Functions

1. Record estimated and actual labor hours expended by maintenance craftsmen. Maintain a history by craft of all labor hours expended on each work order.

2. Provide the ability to delete, change, or add labor hours for each craft for each work order.

3. Provide the ability to delete a work order when retention of the craft labor history is no longer required.

4. Produce the following reports:
   a. Input Transaction Report—record of all input (i.e., batch listing noting all errors and rejections).
   b. Labor Hours by Work Order Number—a listing by Work Order Number of all craft hours expended on the job.
   c. Labor Hours by Craft—a listing of all labor hours expended for each craft against Construction Work Orders.

The Construction Labor Hours System will be run on a weekly basis.

Processing Steps Leading Up to the Construction Labor Hours System

1. The maintenance operations clerk will open the work order in the masterfile using a Work Order Loadsheet.

2. The timekeeping clerk will record the labor hours for each work order from the daily timesheets onto the Labor Activity Loadsheet.

3. The operations clerk will review the timekeeper's loadsheets and batch them for input.

4. When the Work Order History is no longer required, the operations clerk will delete the work order from the masterfile using a Work Order Transaction Loadsheet. This process will normally be accomplished in the month following completion of the work order.

FIGURE 22-3A

SAMPLE SYSTEMS REQUIREMENTS ANALYSIS

SYSTEMS REQUIREMENTS ANALYSIS REPORT

DATA ELEMENTS

	Work Order No.	Craft 1 Carpenters		Craft 2 Painters		Craft 3 Plumbers and Sheet Metal Workers		Craft 4 Electricians	
		Estimated Hours	Actual Hours	Estimated Hours	Actual Hours	Estimated Hours	Actual Hours	Estimated Hours	Actual Hours
Input Documents									
Work Order Loadsheet	X	X		X		X		X	
Labor Activity Loadsheet	X		X		X		X		X
Output Reports									
Input Transaction Report	X	X	X	X	X	X	X	X	X
Labor Hours by Work Order	X	X	X	X	X	X	X	X	X
Detail Labor Hours by Craft	X	X	X	X	X	X	X	X	X
Summary Labor Hours by Craft	X	X	X	X	X	X	X	X	X

FIGURE 22-3B

SAMPLE SYSTEMS REQUIREMENTS ANALYSIS

STEP 3—THE CONCEPTUAL DESIGN AND SPECIFICATIONS PHASE

The conceptual design and specifications phase accomplishes those activities that translate the system requirements developed in Step 2 into a conceptual design with attendant specifications. For the most part, work on this phase is performed by the data processing programmer analysts assigned to the project. The major considerations consist of (1) selecting the computer installation to be used for the automated system, (2) determining the systems flow, and (3) preparing the specifications for source documents and report outputs.

Figure 22-4A is an example of a systems flow diagram for the construction labor hours system described in Figures 22-3A and 22-3B. The diagram indicates the flow of source documents into the system, the punching of cards, and the creation of a transaction file and report, a new masterfile tape, and a report tape for outputs. Note that the computer installation, the program language to be used, and the schedule are also identified. Figure 22-4B presents some of the most common symbols used in flow charts.

Figures 22-5A and 22-5B are examples of specifications for reports from the sample construction labor hours system. The specifications include eight elements: Report Number, Title, Purpose, Report Format, Sort Sequence, Totaling, Page Break Criteria, and Report Schedule. They also include a space for describing additional requirements. These exhibits are the specifications for the first two reports listed in the output reports section of Figure 22-3B.

Despite the fact that the conceptual design and specification phase is largely a matter of applying computer technology, maintenance department personnel should definitely participate. This participation should encompass testing for four criteria:

- Viability
- Timeliness
- Relativity
- Veracity

TESTING FOR VIABILITY

Testing for viability consists of three items that should be covered to support the criterion of viability. First, assurances should be obtained from the data processing personnel assigned to the project that the computer program language and data processing equipment selected for the system will not be rendered obsolete in the foreseeable future. This assurance is necessary because the system cannot handle changes or growth if the computer program is rendered obsolete.

Second, the programmer analysts assigned to the project should be fully aware of any anticipated changes that may have to be made to the computer program. For example, in Figure 22-3B the systems requirements analysis for the construction labor hours system indicates only four different crafts. At some point in the future it might be desirable to add additional codes to designate other crafts. The programmer analyst should be aware of additional craft codes that may be needed so that he can construct the program in a manner that will permit them to be readily added in the future.

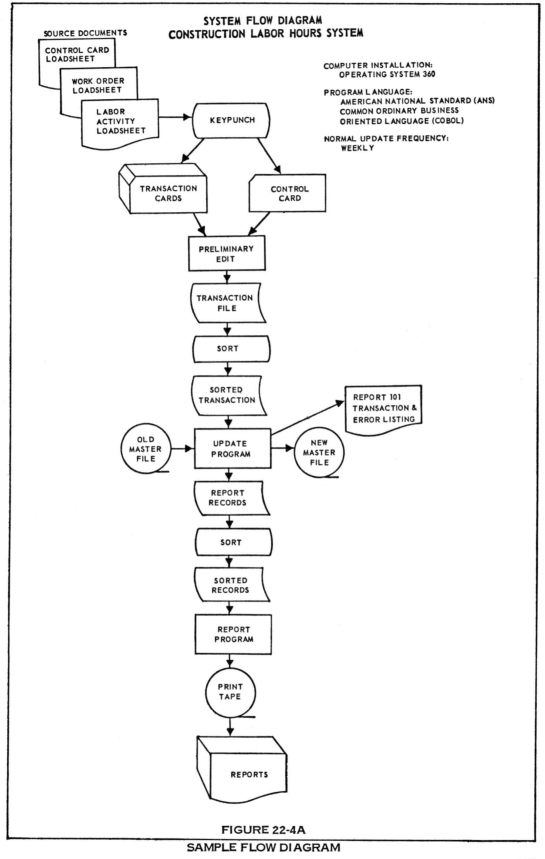

SYSTEM FLOW DIAGRAM
CONSTRUCTION LABOR HOURS SYSTEM

SOURCE DOCUMENTS

CONTROL CARD LOADSHEET

WORK ORDER LOADSHEET

LABOR ACTIVITY LOADSHEET

KEYPUNCH

COMPUTER INSTALLATION:
OPERATING SYSTEM 360

PROGRAM LANGUAGE:
AMERICAN NATIONAL STANDARD (ANS)
COMMON ORDINARY BUSINESS
ORIENTED LANGUAGE (COBOL)

NORMAL UPDATE FREQUENCY:
WEEKLY

TRANSACTION CARDS

CONTROL CARD

PRELIMINARY EDIT

TRANSACTION FILE

SORT

SORTED TRANSACTION

OLD MASTER FILE

UPDATE PROGRAM

NEW MASTER FILE

REPORT 101 TRANSACTION & ERROR LISTING

REPORT RECORDS

SORT

SORTED RECORDS

REPORT PROGRAM

PRINT TAPE

REPORTS

FIGURE 22-4A

SAMPLE FLOW DIAGRAM

SYSTEM FLOW CHART SYMBOLS

The first seven symbols listed in the left column of this exhibit are the flow chart symbols used in Figure 22-4A. Other symbols shown are those which are most frequently used in flow charting computerized data processing systems.

DOCUMENT Paper documents and reports of all varieties.	**ONLINE KEYBOARD** Information supplied to or by a computer utilizing an online device.
FLOW ◁ ▷ ▽ △ The direction of processing or data flow.	**DISPLAY** Information displayed by plotters or video devices.
KEYING OPERATION An operation utilizing a key-driven device.	**SORTING, COLLATING** An operation on sorting or collating equipment.
PUNCHED CARD All varieties of punched cards including stubs.	**AUXILIARY OPERATION** A machine operation supplementing the main processing function.
MAGNETIC TAPE	**COMMUNICATION LINK** The automatic transmission of information from one location to another via communication lines.
PROCESSING A major processing function.	**CLERICAL OPERATION** A manual offline operation not requiring mechanical aid.
DISK, DRUM, RANDOM ACCESS	**OFFLINE STORAGE** Offline storage of either paper, cards, magnetic or perforated tape.
TRANSMITTAL TAPE A proof or adding machine tape or similar batch-control information.	**PERFORATED TAPE** Paper or plastic, chad or chadless.

FIGURE 22-4B

SAMPLE SYMBOLS

278

Report Number: 101

Title: INPUT TRANSACTION AND ERROR LISTING

Purpose: To provide a mirror image of all input additions, changes and deletions made to the masterfile and all rejections during an update run.

Report Format:
This format is to consist of a single line report with column headings reading from left to right as follows:

TRANSACTION CODE, WORK ORDER NUMBER, CRAFT CODE, ESTI-MATED LABOR HOURS, ACTUAL LABOR HOURS, ERROR TYPE

Sort Sequence: Construction Work Order Number, Craft Code, Transaction Code

Totaling:
At the end of the report print out the total number of:

Additions
Changes
Deletions
Rejections

Page Break Criteria:
Print as a straight-line listing except show totaling on a separate page

Report Schedule: Weekly

Additional Requirements: ERROR MESSAGE CODES

The following are the error message codes to appear in the error type column of the Transaction Report to indicate why a given input was rejected. The data that caused the rejection is also to be underlined with asterisks.

Code	Meaning
1	The Work Order Number addition was incorrect. First digit must be alpha, the succeeding 5 digits must be numeric.
2	Craft Code designation was not a value of 1, 2, 3, or 4.
3	Estimated Labor Hours was not numeric.
4	Actual Labor Hours was not numeric.
5	The Transaction Code on the input contained other than A, C, or D.
6	This is a change or deletion to a Work Order Number that does not exist in the masterfile.

FIGURE 22-5A
SAMPLE REPORT SPECIFICATION

CONSTRUCTION LABOR HOURS SYSTEM
REPORT SPECIFICATIONS

Report Number: 102

Title: LABOR HOURS BY WORK ORDER

Purpose: To indicate the weekly and cumulative number of actual labor hours expended by craft against a given work order, the total estimated hours, and the difference between the estimated and cumulative actual hours.

Report Format:
This format is to consist of a single line report with column headings reading from left to right as follows:

CRAFT CODE, WORK ORDER NUMBER, LABOR HOURS, CURRENT WEEK, LABOR HOURS CUMULATIVE, LABOR HOURS ESTIMATE, DIFFERENCE

Sort Sequence: Work Order Number, Craft Code

Totaling:
Current week labor hours by:
    *Work Order Number
    **Report

Page Break Criteria: None

Report Schedule: Weekly

Additional Requirements:
For each Craft Code in which the actual labor hours exceeds the estimate labor hours, precede the value shown in the DIFFERENCE column with a minus sign and underline with asterisks.

FIGURE 22-5B

SAMPLE REPORT SPECIFICATION

Third, the programmer analysts should be advised of growth and volume requirements that should be anticipated in designing the computer program. Otherwise, extensive reprogramming may be required in the future to handle growth.

A multiplant manufacturing firm developed a cost accounting system that was to process cost data on maintenance work orders. The programmer designed the system to receive a maximum of 300 work orders per week. During the implementation phase, it was discovered that the weekly volume was about 4000 maintenance work orders. As a result, the computer program had to be revised to handle this increase in volume above the 300 work orders that it was originally designed to process. This delay in the implementation phase and the required revision to the computer program could have been avoided. The programmer should have been told during the design concept phase that the maintenance work orders entering the system would come from off-site locations as well as the central plant.

TESTING FOR TIMELINESS

The second activity to be performed by maintenance department personnel is a review of the proposed systems flow to assure that it meets the criterion of timeliness. For example, if the maintenance department needs weekly reports, the system should be designed to handle weekly updates. A monthly update cycle would fail to provide the proper timing for the information required.

TESTING FOR RELATIVITY

The third activity of maintenance department personnel is a review of report formats and screen displays for on-line, remote terminal viewers. The data elements, formats, and frequency of the reports should meet the needs of using personnel. This review will revalidate the conditions established in the requirements analysis phase. It will also help to assure that nothing was lost in the translation of these requirements into report or screen display specifications.

TESTING FOR VERACITY

A fourth activity that should be performed by maintenance department personnel tests veracity by accomplishing the following:

1. Identify Required Editing Routines
2. Design Source Documents to:
   - Minimize the number of manual entries
   - Size the document for the user
   - Facilitate input

Identifying Required Editing Routines

Without editing routines, the data in the automated system and the reports that will be received may not be accurate. To the maximum extent practicable, requirements for these checks should be identified during the design concept and specifications phase. Figure 22-6 is an example of editing routines applicable to the sample construction labor hours system.

EDITING ROUTINES FOR THE CONSTRUCTION WORK ORDER CRAFT LABOR SYSTEM

Data Element	Required Edit	Action on Error	Error Message Code Number
1. Work Order Number	The first digit must be alpha. The succeeding five digits must be numeric	Reject entire input. Display on the Transaction Report with the erroneous work order number underlined with asterisks and error message code of 1	1
2. Craft Code	Must be a numeric value of 1, 2, 3, or 4	Reject entire input. Display on the Transaction Report with the erroneous craft code underlined with asterisks and an error message code of 2	2
3. Estimated Labor Hours	All digits must be numeric	Reject entire input. Display on the Transaction Report with the erroneous labor hours underlined with asterisks and an error message code of 3	3
4. Actual Labor Hours	All digits must be numeric	Reject entire input. Display on Transaction Report with the erroneous actual labor hours underlined with asterisks and an error message code of 4	4

FIGURE 22-6

SAMPLE EDITING ROUTINES

Designing Source Documents

Several things should be considered when a design of the proposed formats of the source documents is made. Forms used as source documents for data are filled in by people. These documents are subsequently used by a keypunch or a remote terminal operator to provide input to the automated system. Because the fact that source documents are filled in and processed by people is not given adequate consideration, there is often a lack of human engineering in designing the forms.

Minimizing the Number of Entries

Source documents that are not designed to facilitate the entries to be made manually are distasteful to those who must use them, which not only results in criticism, but also in errors. The number of entries to be made on the form by nonclerical personnel should also be minimized. For example, clerical work is not usually placed in high regard by maintenance craftsmen. If craftsmen are required to make a considerable number of entries on a maintenance work order, they will probably not be too enchanted with the system, and erroneous entries will be made. In addition, there is a cost factor, since craft labor wages are being paid for clerical work.

Figure 22-7 is the format of a maintenance work order that was specifically designed to minimize entries by the craftsmen. Although the form contains thirty blocks for the entry of different types of information, only five of these blocks are filled in by the craftsman. He is simply required to enter his name, the date he performed the work, his cost center code, the time he spent on the job, and any pertinent repair and parts replacement information.

Sizing Source Documents

The size of the form should also be considered. The maintenance department of an air frame manufacturer designed a computer-generated scheduled maintenance work order that was eleven inches wide and fourteen inches long. This size of paper was not easy to handle in the field. As a result, the forms were often damaged and rendered partially illegible from being carried around in the craftsman's tool box. Another maintenance department did not experience these problems because their form was specifically designed to fit into the craftsman's shirt pocket. The form was easy to handle in the field and was protected from damage. See Figure 22-7.

Facilitating Input

A form that is not designed to facilitate use by a keypunch or remote terminal operator also can cause errors and an increase in operational costs. Keypunch operators prefer to read and punch data moving in sequence from left to right and down the source document the same way we read a book. When a keypunch operator has to play hopscotch to punch data from a form, the probability of errors increases sharply. In addition, a difficult keypunching job takes more time for the operator to perform and will increase the total costs of operating the system because of the increase in keypunching costs.

Figure 22-8A is an example of a maintenance work order that was not designed to facilitate keypunching. Note how the circled numbers that represent the punching order for the data require the keypunch operator to skip all over the form to punch the information in the required sequence.

Figure 22-8B represents a revision to the form to support keypunching operations. With the exception of the third data element (work order number), the redesigned form permits the keypunch operator to move from left to right and down the form in punching the data. Figures 22-8A and 22-8B are actual forms used by the maintenance department of a Southern California manufacturing firm. The revision to the form reduced both keypunching time and the number of errors.

MAINTENANCE WORK ORDER   NO. 358781

KEY NUMBER	OPERATION NUMBER	REQ. CCC	JOB NO.	ID NO.	LOCATION BLDG.	ROOM	OPERATION DESCRIPTION	NOMENCLATURE	FUNCTION	CRAFT CODE	SCHED. WEEK	T W

REQUESTER   EXT.   DATE

REMARKS:

M.R. NO.	NAME	DATE	CCC	HOURS
	XXXXXXXXXXXXXXXXXXXXXXXXXXXXXXXXXXXXX			

TOTAL HOURS

F.W.O.NO.

M.R. TOTAL COST   F.W.O. TOTAL COST

FOREMAN	DATE COMP.	WEEK COMP.

REPAIR DATA   XXXXXXXXXXXXXXXXXXXXXXXXXXXXXXXXXXXXXXXXXXXXXXXXXXX

For the above Maintenance Work Order, mandatory data blocks to be filled in by craftsmen are NAME, DATE, CCC (Cost Center Code) and HOURS (to the nearest tenth of an hour). When appropriate, the REPAIR DATA block is also to be used by the craftsman to enter any pertinent information about repairs made and parts replaced.

FIGURE 22-7
SAMPLE WORK ORDER

④					①	②					SCHED.	TYPE
COST CENTER	BLDG. LOCATION	ROOM	JOB NO.	ID	MACHINE NAME	MACHINE NUMBER	PART OR OPERATION NO.	LUBE/EQUIP.	FUNCTION	CRAFT	WEEK	WORK
REQUESTER	BADGE NO.		EXT.	DATE	M.R. NO.		NAME		DATE	CCC	HOURS	

REMARKS:

AMOUNT OF OIL
GALS. 10ths

FOREMAN — TOTAL HOURS

F.W.O. NO. — M.R. TOTAL COST

REPAIR DATA — F.W.O. TOTAL COST

| DATE COMP: | WEEK COMP: | ERROR CODE: | TOLER-ANCE: |

MAINTENANCE WORK ORDER

③ 58006

Data elements ①②and③are not in left to right sequence, so the keypunch operator must begin punching in the middle of the first line of the form to pick up data elements①and②, skip to the lower left-hand corner for data element③, and skip back to the top line on the form to pick up data element④.

FIGURE 22-8A
SAMPLE ORIGINAL WORK ORDER

MAINTENANCE WORK ORDER

NO. 358781 ③

KEY NUMBER	OPERATION NUMBER	REQ. CCC	JOB NO.	ID NO.	LOCATION BLDG.	ROOM	NOMENCLATURE	OPERATION DESCRIPTION	FUNCTION	CRAFT CODE	SCHED. WEEK	T W
①	②	④										

REQUESTER		EXT.		DATE							

REMARKS:

M.R. NO.			NAME		DATE	CCC	HOURS

TOTAL HOURS

F.W.O. NO.	M.R. TOTAL COST	F.W.O. TOTAL COST

	FOREMAN	DATE COMP.	WEEK COMP.

REPAIR DATA

In this revision of the format shown in Figure 22-8A data elements ① and ② have been placed to the left of data element ④ on the first line of the form. Data element ③, the preprinted work order number, has been moved to the top of the form. These changes were made to facilitate the keypunching operation.

FIGURE 22-8B
SAMPLE REVISED WORK ORDER

STEP 4—THE PROGRAMMING AND TESTING PHASE

The actual computer programming for an automated system is performed by the programmer analysts assigned to the project. This, however, does not mean that maintenance department personnel have no role to play during the programming and systems test phase. Their role should consist of the following four activities:

- Assuring Relativity
- Preparing Test Data
- Testing Source Documents
- Assuring Viability

ASSURING RELATIVITY

While the programming is being accomplished, questions will probably arise regarding systems requirements that require clarification. The maintenance manager should be prepared to assure that answers to these questions are provided in a timely manner by appropriate maintenance personnel. If the answers are not provided, the programmer analysts are forced into the position of having to make their own arbitrary decisions. This situation can lead to a marked difference between what was wanted from the automated system and what is provided.

PREPARING TEST DATA

The second activity that should be performed by maintenance department personnel is the preparation of test data for the computer program. There are two reasons for having maintenance personnel perform this task. The first reason is that maintenance personnel are more familiar with the information and its content than the programmer analysts who are assigned to the project, so the test data will be closer to the actual information that is to be processed by the system. The second reason is that the preparation of the test data by maintenance personnel provides the opportunity for training in the preparation and transmittal of source documents to be used in the automated system. In preparing test data the following criteria should be met.

1. The test data should contain several samples of each source document used in the system.
2. If there are editing routines in the program, the test data should be designed to include sample errors that will test the proficiency of the editing routines. The test data should contain several examples of each error that is to be processed by each editing routine.
3. There should be sufficient samples of data for each field in the file to provide enough examples to test all report writing routines with respect to sorting, totaling, and page break criteria.

TESTING SOURCE DOCUMENTS

The testing phase not only provides the opportunity for testing the adequacy of the computer program and the training of personnel, it also permits the testing of the source documents to be used in the system. There is nothing better than the actual use of a form by people to determine if adequate human engineering has been applied in its design.

There have been occasions when large quantities of forms have been obtained before they have been tried out in the test phase. These large quantities are purchased in order to amortize the costs of art work, plates, and setup over a large number of forms, and thereby reduce the unit cost of each form. However, if the forms prove to be less than adequate during the test phase, two alternatives are available. Either the existing forms have to be scrapped, or the necessary revisions are not made until the form is reordered and people are forced to use up the existing stocks. Neither of these alternatives is desirable. Scrapping the existing forms and replacing them incurs additional costs. Waiting until the reorder point to make the revisions usually results in more errors and more time being expended in preparing and transmitting the data. This additional labor also results in increased costs. To avoid this problem, only limited quantities of forms should be purchased until the source documents have been tried out in the test phase.

ASSURING VIABILITY

Assuring viability is largely a matter of the logic applied in preparing the computer program and the manner in which the information masterfile is structured. It is largely up to the programmer analysts to build in viability during the programming phase. One way maintenance department personnel can assist them is by identifying data elements that may be subject to frequent change, such as building numbers, cost center codes, charge-out rates, and standard labor hours. This maintenance department help will facilitate the development of an efficient computer program for the maintenance of this type of data. Assistance can also be rendered by reexamination of the anticipated growth of the system to assure that the information compiled in the design concept phase is still valid.

STEP 5—THE IMPLEMENTATION PHASE

The implementation phase consists of placing the automated system in an operational mode. There are two basic techniques for accomplishing this phase. One method is the "parallel" approach, which consists of running the old and the new systems simultaneously for a prescribed period, with a gradual change-over. This method is normally used in an effort to provide a smoother transition from the old to the new system. The other technique is the "cut-off" approach, which consists of making a complete change-over on a prescribed date. This method is normally applied where both the old and the new systems are automated. It has the advantage of immediately eliminating the costs of running the old system.

Regardless of which technique is utilized, the implementation phase should consist of the following three activities:

- Establishment of the initial masterfile
- Training of personnel
- Completion of systems documentation

ESTABLISHING THE MASTERFILE

This activity requires the preparation and input of all source documents for the initial information to be contained in the automated information system. Because this activity can involve a substantial amount of clerical effort on the part of maintenance personnel, maintenance supervisors should carefully plan for this effort.

TRAINING

Training of personnel includes the training of maintenance personnel who will use the system and data processing personnel who will operate the system. No matter how sophisticated it is, an automated information system is only as good as the information fed into it. If maintenance personnel are not adequately trained to make proper inputs, the data submitted to the system will invariably contain a certain amount of garbage. In turn, the output from the system, in the form of reports, will also reflect this garbage. As a result, the reliability of the system is placed in jeopardy. Through the application of the veracity criterion in the programming phase, some of this garbage can be rejected by the computer program. However, rejected data must be corrected and resubmitted, which means additional clerical costs and a delay in having the information available in the automated information system. In other instances, the editing routines may be programmed to automatically correct the data prior to processing. Extensive editing routines, however, are costly because of the additional programming effort required to develop sophisticated routines.

In addition, computer time is required to screen the input data through the editing routines. This additional amount of computer time required to process the data also increases the total cost of operating the system. Thus, editing routines are beneficial, but they are not a complete substitute for adequately trained personnel. They are only effective in rejecting or correcting the errors in the data submitted and cannot replace trained personnel.

Back-up personnel should be trained as well as those members of the maintenance department whose primary duties involve the use of the automated system. A frequently encountered operational problem is the lack of adequately trained back-up personnel. Maintenance management should not rely on one person to maintain the automated information system. Alternate personnel should be trained in the preparation of source documents and the batch processing routines or the operation of the remote terminal. Otherwise, when the person who normally performs these functions is not available, the whole system stops, or the person who serves as a substitute does a poor job due to a lack of training.

DOCUMENTING THE SYSTEM

Documenting the system includes both the documentation of the system that is used by data processing personnel and the system that is used by maintenance personnel. The documentation for the maintenance department is generally referred to as a User's Manual. Prior to publication, the draft of the User's Manual should be thoroughly reviewed by all concerned personnel in the maintenance department for clarity, accuracy, and completeness. If this review is not accomplished, the User's Manual may prove to be less than adequate as a reference guide, particularly when it is used to train new maintenance personnel in how to use the system.

STEP 6—THE EVALUATION PHASE

The final phase in the development of an automated information system consists of reviewing the system after it has become operational. The first review should not normally begin until the system has been fully operational for at least three months. Subsequent evaluations should occur annually.

One facet of an evaluation phase consists of discovering all the things that are wrong with the system. The number of things discovered to be wrong is a measurement of how diligently the criteria of cost effectiveness, veracity, viability, relativity, and timeliness were applied in the preceding five phases. There is, however, a more positive aspect to the evaluation phase, which involves systems enhancements. Even for the most successful system, there will be a need for refinements. Editorial routines will require changes and updating. Revisions to existing reports, the generation of additional reports, and the addition of data elements should be expected as enhancements to the initial system. This is why the application of the criterion of viability is so important in the conceptual design and specification phase and the subsequent programming phase. If viability has not been designed into the computer program, the opportunities for evolutionary enhancements will be limited because of the additional costs for extensive reprogramming.

An adequate evaluation should encompass the following four activities:

- Determination of the error rejection rate on data input
- Analysis of source documents
- Examination of data elements
- Review of report outputs

DETERMINING THE ERROR REJECTION RATE

An error rejection rate is normally expressed as a percentage. The percentage is obtained by dividing the total volume of inputs into the total number of rejections appearing on the transaction reports for a specific period of time. The error rejection rate is used as a measurement of how efficiently data is being prepared and transmitted into the automated system. As a rule, an acceptable error rate may be considered to be less than 2 percent. Once the rate is determined, it is necessary to discover where and why these errors are occurring. The two potential sources of errors are the people who are filling in the source documents and the people who are keypunching the data or transmitting it through a remote terminal. Examination of the source documents from which the data was taken will reveal if the error was clerical. If the data on the source document is correct, the error occurred in the transmittal.

One cause of errors may be a lack of knowledge on the part of the personnel preparing or transmitting the data. This problem can be remedied by additional training. A second cause of errors may be a lack of clarity in the instruction manuals. This problem should be rectified by revising the instructions. A third cause of errors may be the design of the source document. This problem leads us to the next activity to be discussed.

ANALYZING SOURCE DOCUMENTS

The analysis of source documents consists of reviewing the source documents for their adequacy from the standpoint of users. Figures 22-8A and 22-8B show the changes in the format of a maintenance work order that were made to facilitate the keypunch operation. These changes were made as the result of a systems evaluation that occurred approximately two years after the system was fully operational. Note that in the center of Figure 22-8A there is a data block entitled "Amount of Oil," and that in the lower right-hand corner there are two blocks entitled "Error Code" and "Tolerance." None of these blocks appear on the form in Figure 22-8B. The fact that these data elements were removed leads us to a discussion of the next activity.

EXAMINING DATA ELEMENTS

As previously mentioned, a systems evaluation may reveal that certain data elements should be added to the system to enhance its reporting capabilities. Conversely, an evaluation may also identify data elements already in the system that are superfluous. Figures 22-8A and 22-8B demonstrate this point. The data elements "Amount of Oil," "Error Code," and "Tolerance" were eliminated as active fields in the masterfile for this maintenance system when an evaluation revealed that these data elements were no longer being used. The redesigned form shown in Figure 22-8B reduced keypunching costs by 3 percent, resulting in an annual savings in excess of $1000 per year, and reduced keypunching errors by greater than 1 percent.

REVIEWING REPORT OUTPUTS

The reviewing of report outputs entails two considerations. First, there may be changes desired in existing reports or screen displays for on-line systems. These may include the addition or deletion of data elements or revisions to the sort sequence of the data elements, the totaling routines, the page breaks, and the report frequency. Second, there are additional reports or screen displays that may be desired from the system. If it has been determined that data elements are to be added to the system, then, invariably, existing reports will have to be changed or additional reports generated in order to make available the stored information for these added data elements.

ACQUIRING EXISTING SOFTWARE

There are some computer programs for maintenance operations that are commercially available on either a lease or buy arrangement. In a sense these canned packages are existing solutions looking for problems. As such, the software will probably not be as perfect as a tailor-made computer program specifically designed to meet the needs of a specific maintenance department.

On the other hand, a maintenance department may not be able to afford the luxury in developmental costs that a tailor-made system entails. Further, if the package has been developed as a solution to the problems of several clients, it may even be more sophisticated than the department's proposed system. Thus, a commercially available program may be adequate for the maintenance department's needs.

The main advantages in buying a package are (1) a savings in developmental costs and (2) reduced implementation time. The first advantage is a definite cost avoidance for the organization. The second advantage may constitute hard savings depending upon how well the system achieves cost reductions n maintenance operations.

DETERMINING SUITABILITY

The analysis of the suitability of an existing software package should essentially follow the same steps as the development of a new system. Step 1, the operations analysis phase, should first be performed to determine the need for the automated system. Step 2, the requirements analysis phase, and Step 3, the design concept and specifications phase, should then be performed. The requirements developed should then be compared with the existing software package to see how well there is a match.

It is essential that Steps 1, 2, and 3 be followed correctly and completely, because a decision as to whether or not to acquire the system is normally required upon the completion of Step 3. It is seldom that a vendor will be willing to proceed into Step 4, the programming and testing phase, until there has been a contractual commitment for the lease or purchase of the computer program.

TESTING

Step 4, the programming and testing phase, takes on a somewhat different dimension with respect to a canned package. The computer program already exists. There may be some adjustments to tables or other stored data that is changeable in the program. However, in the main, this phase consists of preparing your test data and seeing how well the computer program works.

It is essential that all testing be accomplished prior to any implementation. Normally, a vendor supplies the testing as part of the sales price. However, the vendor normally charges for any additional work that may be required after completion of the testing phase. So if the testing is inadequate and bugs are discovered in the implementation phase, you will pay extra to get them worked out.

IMPLEMENTING

After testing, the implementation phase for a canned package is essentially the same as for a developed system. The main difference is that the vendor will have already delivered the documentation. You should plan on having to make some revisions in the User's Manual to make it more suitable to your particular needs.

As with internally developed systems, changes in the canned package should be anticipated. Viability should be considered in whatever arrangements are made to acquire the rights to use the commercial package. Make sure that the firm that sells the package either guarantees to provide this service or provides your organization with a source deck and sufficient documentation so that changes can be performed in-house.

If the price is right, the acquisition of an existing software package for use on an interim basis is an excellent method of solving your immediate needs for an automated system while you develop your requirements for a more sophisticated, tailor-made system. The canned package may show what type of tailor-made system is preferable. One university acquired a canned package for a periodic maintenance system for about $25,000, and used it for three years. The purchased program gave them what they initially needed to establish an automated periodic maintenance system. It saved them some $40,000 annually in clerical labor. After three year's experience, they were able to determine what they needed in a better system, which would interface with their accounting system and could provide more information through additional data elements. They proceeded to develop their own system over the next two years. The canned package operated for a total of five years. The $25,000 cost yielded a net savings in clerical labor of about $175,000 and provided experience in the use of a computer-based information system.

SECTION V–DIRECTING

Food for thought:

FOR THE SUBORDINATE

The wise man is glad to be instructed, but a self-sufficient fool falls flat on his face.

To learn you must want to be taught. To refuse reproof is stupid.

FOR THE SUPERIOR

Be patient and you will finally win, for a soft tongue can break hard bones.

Timely advice is as lovely as golden apples in a silver basket.

FOR SUPERIORS AND SUBORDINATES

The intelligent man is always open to new ideas. In fact, he looks for them.

If you wait for the perfect time, you will never get anything done.

On with it then, and finish the job! Be as eager as you were to plan it, and do it with whatever you have.

YOU ARE NOT RAISING MUSHROOMS

To have good mushrooms, you have to keep them in the dark and feed them lots of manure. But maintenance personnel are not mushrooms and this process is not recommended for use in directing a maintenance department. In fact, successful direction requires the exact opposite process, continuous orientation and honest communication by the manager.

POINTS TO CONSIDER

Direction by the management consists of guiding and supervising subordinates within the maintenance department. To direct subordinates, a manager must communicate, motivate, and lead. No two managers will do these things exactly the same, because no two managers are the same and no two managers have the same group of individual subordinates to direct.

Directing is a complex job. It is complex because we are dealing exclusively with the human factor and all those actions that are designed to encourage subordinates to work efficiently and effectively both in the short run and the long run. The human factor is unlike the inanimate land and capital factors of production. Subordinates are not a purchased commodity. An individual is complex, with a long series of often changing needs. The maintenance manager's challenge is to evoke the productive power of each individual that is required by the organization. The manager is responsible for exercising the care of subordinates that encourages them to contribute fully to the department's objectives.

Behavorial scientists, management theorists, and academicians have written volumes on the subjects of leadership, communication, and motivation. In these writings there are a variety of claims of truth, diverse conclusions, and some plain gobbledygook.

This section of the book examines some of the valid fundamentals and demonstrates some proven methods of executing them. The first chapter presents some of the principles of direction. The second chapter covers how and when to use them.

23 *Guidelines for Directing Maintenance Supervisors*

A successful management team in a maintenance department is built on a foundation of confidence and understanding that withstands the tense pressures and demands that tight schedules and inflexible deadlines invariably create. This essential foundation of confidence and understanding is mainly created through careful direction by management.

This chapter is important because it covers the basic principles that must be applied for successful direction. How and when to apply these principles is covered in chapter 24.

To direct subordinates, a manager must motivate, communicate, and lead. All the work that has gone into organizing and staffing can be nullified, and the objectives called for by plans never attained, if direction is faulty. Although many methods and devices are available for use in achieving control, there is still the need for personal observation. Budgets, charts, and all the other control techniques may be helpful, but you can't just sit and review data in your office and expect to achieve control. There is always a need for the direction of people if you want them to do what has been planned. Let's look at an example.

The manager of a facilities maintenance department decided to take a look at how the second-shift operations were getting along. As he toured the carpenter shop, there was a man cutting plywood remnants into smaller pieces with a table saw. When he asked the carpenter what he was doing, the man replied: "I cut these pieces so that we can get rid of them as scrap. The foreman don't want a bunch of remnants around cluttering up the shop."

The matter was handled properly. The manager did not issue orders directly to the carpenter. After the tour, he had a chat with the foreman about conservation. The foreman, in turn, changed the procedures and established a nesting gain operation for sheet stock. Dealing with the foreman rather than with the carpenter was proper execution under the principle of unity of command. This principle is discussed in more detail later in this chapter.

This example also provides an object lesson in priorities. The foreman put the highest priority on housekeeping and a neat looking shop. The manager was more concerned about cost-effective utilization of resources, both with respect to materials and the time spent by a carpenter in destroying the remnants. Orienting subordinates and meshing objectives by establishing priorities between various goals are essentials in direction by management.

MOTIVATING

There are many diverse opinions about motivation. At best, the present state of knowledge on this subject is a long way from being adequate. Let's take a pragmatic approach and talk about what motivation is supposed to do, rather than dwell upon how it works.

EVALUATING A MOTIVATIONAL SYSTEM

To motivate means to provide with a motive. A motive is something that causes a person to act and implies an emotion or desire operating on the will and causing it to act. Some theorists claim that the motivation system must provide both positive and negative incentives and meet the ego and self-development needs of subordinates if it is to create a desire in people to work. Other theorists claim that inducements and goals do not truly motivate, since work will soon become a drudgery if it is not challenging. For them, true motivation requires that a person have job enhancement or fulfillment to develop the right attitude and put out the needed effort.

Regardless of which theory suits you, it should be recognized that the responsibility for motivation is definitely restricted by the amount of delegated authority within an organization. A maintenance manager is limited by the parent organization's policy on salaries, fringe benefits, promotion, opportunities for job enhancement, and the extent of centralization of authority. He can only work within the confines of his authority. However, within the confines laid upon him, the maintenance manager is responsible for motivating not only his immediate subordinates, but all subordinates right down to the worker level.

The acid test of any motivational system is whether or not it is productive in the sense of having subordinates who work efficiently and effectively. Financial opportunities, fringe benefits, and opportunities for desired advancement and job fulfillment are meaningless if subordinates are not producing. Adequate motivation must satisfy needs and release the capacity for work.

When we finish wading through all the theories, we discover that the best motivators are managers who understand their fellow persons and who follow their native instincts in dealing with them. They gauge their methods and the extent of direction to each individual subordinate.

SIGHTING IN

It may sound trite, but there are two tried and true guidelines for motivating. These are:

1. **The biblical admonishment to do unto others as you would have them do unto you.**

If the golden rule were truly followed, any system of eliciting effort would be headed in the right direction.

2. **Honesty is the best policy.**

When Ben Franklin coined this phrase, he was not just moralizing. He was basing this principle on material interest as the method of action that would yield the highest results in the long run.

The ability of a manager to motivate can be evaluated to a marked degree by his ability to create integrity and loyalty in subordinates. The creation of integrity and loyalty is brought about by instilling values, which, in turn, create attitudes. If the manager does not practice and insist upon a high standard of business ethics, there is little hope of his creating integrity or loyalty.

COORDINATING OBJECTIVES

The central task of a manager is to reconcile differences in approach, effort, timing, or interest, and *to synchronize individual and cooperative goals.* Effective direction depends on the extent to which individual and departmental objectives are synchronized.

It should be recognized that individuals who are employed in a maintenance department have personal goals that may be different from the goals of the department or the parent organization. If the department is going to get anywhere, the goals of individuals have to complement the department's goals. Otherwise, the individual will not make the contribution to departmental goals that he or she should be making.

To coordinate goals, you have to know your subordinates' goals as well as your own for the maintenance department. In directing subordinates, you must take advantage of individual motives to gain departmental goals. In interpreting plans and job assignments, you must harmonize individual and group objectives. It is naive to assume that selfless devotion exists in many, if any, subordinates, or to expect the goals of your subordinates to be identical to the goals of the department. Few people actually identify their goals with the goals of the organization. A "company" man may make your job as a manager easier, but he is a rare person.

The coordination of objectives is largely achieved through a motivation system that enables employees to attain their ego needs through rewards for creativity or innovation, their social needs through group efforts in the work situation, and their basic needs through adequate wages and fringe benefits.

STAFFING

It should also be recognized that there is a direct relationship between staffing and motivating. The more a subordinate is qualified and trained, the more productive his effort is. Proper staffing reduces the costs and needs for a motivational system.

COMMUNICATING

Communication is the means by which employees in a maintenance department exchange information concerning operational requirements, the status of projects, and ideas for improving operational efficiency. Effective communication requires that either the verbal or written message be expressed in commonly understood language.

Badly expressed messages cause a need for follow-up clarification, as well as faulty translations and unclarified assumptions.

Maintenance managers have a significant need for effective communications because a large part of the work force is often geographically scattered during the workday performing field operations. Unlike in an office operation, you cannot get them all back together again readily to provide follow-up clarification, nor can you readily detect faulty translations or assumptions made in the field.

LEADING

There are certain concepts regarding leadership that you should try to teach to your subordinates. These are:

- Styling leadership
- Building confidence
- Creating enthusiasm

These techniques involve trying to teach a lot to a subordinate. It is probably just as hard to practice them yourself. But let's take a look at them.

STYLING LEADERSHIP

In a maintenance department, there is often the need for quick action due to emergency situations. A manager must be authoritative, because there is no time for the use of a participative approach. In other instances, such as planning the annual budget or using the management by objectives approach in the performance appraisal of subordinates, there is time for participation. Each manager has to use those techniques that get superior performance. There is no one best way.

It is currently in vogue in some circles to view a manager's leadership style in terms of a continuum. As indicated in Figure 23-1, an autocratic style is on one end of the continuum, the democratic approach in the middle, and the free-rein style on the other end.

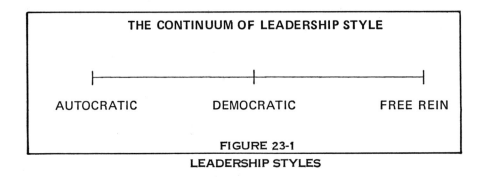

THE CONTINUUM OF LEADERSHIP STYLE

AUTOCRATIC   DEMOCRATIC   FREE REIN

FIGURE 23-1
LEADERSHIP STYLES

The autocratic approach relies on the manager's authority to command subordinates. In using the democratic technique, the manager practices participative management—he solicits advice and may even accept the majority vote of subordinates in reaching decisions. Use of the free-rein technique involves a deliberate abstention from direction or interference with the subordinate's freedom of choice and action. The manager becomes a mild consultant and may even go so far as to abdicate his decision-making authority.

Actually, there is no room in maintenance management for a person who might fall into the classification at either extreme of the continuum.

The complete autocrat is virtually an extinct species. He only survives as an individual proprietor. One problem involved in having this type of person in your management team is that he will never keep any subordinates under him who have a potential for future advancement or development. He will only tolerate yes-men. Another problem is that the sandpaper personality usually associated with the complete autocrat is not effective in dealing with workers or in interacting with peers.

The mild consultant who has abdicated his decision-making authority is at the furthest point on the free-rein end of the scale. This abdication of decision-making authority cannot be tolerated in a maintenance department. Even the most permissive manager should not abdicate his decision-making prerogative as a manager—just as he should not try to divest himself of responsibility by delegating. If he is uncertain about how a decision will affect some of his people, then he should be taught to find out before he makes the decision. If he needs information from his subordinates in order to make a decision, he should be taught to get it so he won't delegate decision making to the subordinate who has the data.

Some theorists believe that, depending upon the situation, each manager can and does readily shift his style from free rein, to democratic, to autocratic, as each seems appropriate. In reality, this is seldom the case. Each manager usually adopts a style that suits his personality, and he is not apt to deviate much from it. He will use the style that is most effective for himself, irrespective of whether it is authoritarian, permissive, or in between. This fact should be recognized when reviewing the management style of each subordinate. Don't monkey with it as long as it works effectively, because you probably won't change it any more than you can expect to change his personality.

No manager should assume that all of his subordinates like to, want to, or can participate in decision making. If a subordinate wants to try a democratic style, this possible reluctance should be pointed out to him. He should be prepared for the fact that he may find a subordinate who simply will not respond effectively to a participative management style.

No manager should try to pretend that he is not deciding when he has the responsibility for making a decision. Subordinates will pick this up quickly. In the main, subordinates want purposeful, clear, and effective leadership. Usually, they readily detect indecision, incompetence, or a yearning to be popular. They don't like these qualities because they adversely affect pride, production, accomplishments, and growth.

BUILDING CONFIDENCE

The confidence of a subordinate is dependent upon his sense of security and the quality of his knowledge. A subordinate's confidence in his ability to manage is developed through orientation, supervision, and the promotion of a feeling of security in his position.

To a large degree, leadership involves effectively representing a group before superior levels of authority and responding to the needs of subordinates. If a manager does not go to bat for his people or he is lethargic in reacting to the needs of his subordinates, he is not likely to build much confidence.

CREATING ENTHUSIASM

There is a definite relationship between leading and motivating. Successful leaders use the motivational system by placing emphasis upon opportunities to feed self-esteem, chances to do special assignments, respect for unusual abilities, or pride in belonging to a well-run operation.

The ability to create enthusiasm in employees is largely a matter of the manager's personal attributes and of his judgment in making use of appropriate occasions. Inspirational techniques can be described but not taught. No two managers will use the same words or actions to create enthusiasm in subordinates, because the power to inspire zeal is a personal quality. Subordinates should be taught to develop their own techniques, essentially by trial and error.

TEACHING DELEGATION

Where size or complexity forces delegation of authority, your subordinates must realize—even if you must go out of your way to teach them—that there is a principle of comparative managerial advantage or value. The essence of this principle is that a manager will improve his contribution to the organization if he concentrates on tasks that contribute the most to the maintenance department objectives and assigns other tasks to subordinates, even though he could accomplish the latter better himself. This principle is easy to state and hard to practice. But failure to do it defeats the very purpose of delegation.

LETTING GO

To have effective delegation of authority, you must be willing to release decision-making power to your subordinates. A noted shortcoming of some managers as they come up from the ranks is that they continue to want to make decisions for their previous positions. They are still confirming the hiring of every journeyman or clerk. Doing this takes time and attention away from more important matters. Let's look at an example.

In a facilities department in Southern California, the foreman of the paint shop was promoted to general foreman in charge of construction. But he still tried to run the paint shop in addition to performing his new job. The maintenance manager had to do some hard coaching to finally make him let go. The manager put it this way: "When I put Bill in charge of construction, I promoted Pete to take over as paint shop foreman. I expected that Bill would be available to help Pete get a handle on his new job, but I did not want Bill to keep running the paint shop. Bill was going to have plenty to do if he was going to be effective as general foreman of construction and that was the job he was being paid to do. Bill had to let go of the paint shop for another reason. If he continued to run it along with doing his new job, Pete would never grow into a full-fledged foreman. I don't need yes-men and expect my foremen to wear long pants. If Pete never developed, we would be stuck on two counts. First, we would have a general foreman who would be spending time on the details of running the paint shop instead of concentrating on getting the construction work done. Second, we would be paying a man to be foreman of the paint shop who wasn't really running the operation."

LETTING OTHERS MAKE MISTAKES

No responsible manager will sit idly by and let a subordinate make a mistake which might endanger his position or get the department into trouble. However, continual checking to assure that the subordinate makes no mistakes makes effective delegation impossible. Every subordinate will make some mistakes. He should be allowed to make them and you should be prepared to charge-off their cost as an investment in his development. A series of repeated mistakes can be avoided without stifling development by:

1. Asking discerning questions
2. Counseling patiently
3. Explaining objectives and policies carefully

SETTING THE STAGE

Effective delegation starts with the creation of an atmosphere of receptiveness. Any manager who delegates must be willing to give the ideas of other people a chance. Decisions normally involve some individual discretion. A subordinate's decision may not be exactly the same as the one you would have made. But as long as the decision was essentially the same and achieved the same results, there should be no challenge. Some highly successful maintenance managers not only use the ideas of others, but also have the knack of implanting their ideas in the minds of others and then congratulating them on their ingenuity. This procedure may be considered a sort of gamesmanship, but it is a subtle method of the "sell don't tell" approach to getting subordinates moving.

DELEGATING EFFECTIVELY

There are two techniques that definitely increase the effectiveness of maintenance management. The technique of completed staff work applies to the superior's use of his subordinates as staff. The technique of exception applies to the superior's use of his subordinates as managers. In practice, the two principles are usually interrelated.

Most people in management are alternately managers and members of staff. A person is seldom exclusively one or the other. Usually a person is staff to his superior and manager to his subordinates. Consequently, both the technique of exception and the technique of completed staff work may be used by the same person. These techniques are particularly effective in the maintenance manager's relationship with his immediate superior and in large-scale maintenance operations with more than four levels of supervision.

COMPLETED STAFF WORK

Completed staff work consists of the study of a challenge and the presentation of a solution by the staff in such form that all that remains to be done on the part of the superior is to indicate his approval or disapproval of the proposed action. It is the staff man's job to advise his superior what he ought to do after the staff man has evolved a single proposed action from a thorough study of the matter. This single proposed action should be worked out in finished form. It is the duty of the staff man to work out the details. The superior should *not* be consulted in the determination of the details.

The technique of completed staff work does not preclude a rough draft, but the rough draft must not be a half-baked idea or a presentation of the challenge in piecemeal fashion. It must be complete in every respect except for neatness and the number of required copies. A rough draft must not be used to shift the burden of formulating the solution to the superior. The superior should get answers, not questions.

In operation, the technique of completed staff work means that the superior makes the decisions and the staff does the work. The proper execution of this technique also imposes responsibilities on the superior as well as the staff. When making the assignment, the superior must clearly state what the challenge is and what the staff is expected to do about it. There are no solutions to unknown challenges. The superior should also contribute his experience by telling the staff what he has learned or thinks that he has learned about the situation. The superior should set a target date and be accessible for legitimate progress reports. This system will eliminate getting today's answers to yesterday's challenge or yesterday's answer to today's challenge. It is the staff's job to furnish proposed solutions—not challenges—and the superior's job to evaluate the proposed solutions and make decisions.

TECHNIQUE OF EXCEPTION

As previously stated, the technique of exception is a management technique applicable to the superior's use of his subordinates as managers rather than as staff. Inherent in the application of this technique is a delegation of authority by the superior. In operation, this technique requires that a subordinate manager refer matters to his superior for a decision only when he does not have the authority to make the decision himself. Only exceptional matters are referred upwards to the superior.

The proper execution of this technique involves obligations for both the superior and the subordinate. The superior is not only required to delegate authority but, in addition, must clearly define the authority granted to his subordinate so that it is distinctly understood what decisions fall within the subordinate's jurisdiction. On the other hand, the subordinate should refer to his superior only those matters that extend past the scope of his authority. Further, following the dictates of completed staff work, the subordinate should refer the matter to his superior with a proposed solution.

The manager who properly delegates authority to his subordinates and practices the principle of exception within his organization will relieve himself of the burden of making many trivial decisions.

MAKING THE TECHNIQUES WORK

These deceptively simple techniques must be completely understood before they can become standard operating procedure. It takes real leadership to build them into a maintenance organization.

The practice of the technique of completed staff work requires the superior to steadfastly resist the temptation to do the staff's or subordinate's thinking. He should give them guidance and background information, *but make sure they do their own thinking.*

The practice of the technique of exception requires the superior to clearly define the authority that has been delegated to his subordinate managers. He must see to it that subordinate managers use this authority.

GETTING THE BENEFITS

These management techniques may not solve every challenge, but they will do these things:

1. Free the superior from petty day-to-day challenges, unimportant details, half-baked ideas, voluminous memoranda, and immature oral presentations
2. Multiply the effectiveness of supervision
3. Increase organizational efficiency
4. Focus the superior's talent on decision making rather than day-to-day problem working

One maintenance manager who tries to use these techniques puts it this way: "I use the technique of exception and give my boss completed staff work, because it is the most efficient way to operate. I don't take up a lot of his time, and it makes it a lot easier to get things done for my people and to sell our programs. My people give me completed staff work and practice the exception principle, because it has proven to be one of the best ways to develop them into full-fledged managers in their own right. I couldn't run a three-shift, twenty-four-hour, seven-day-week operation like this one effectively without delegating some authority. The main thing I ask my people to do is to always make sure they report anything that may get the department, myself, or our boss into trouble. Then we try to hit the problems, not the people."

COORDINATING GOALS

The purpose of coordination of goals is to gear individual efforts towards the accomplishment of departmental goals. Coordination is the essence of management and is necessary to accomplish any of the five functions of management. But coordination becomes particularly significant when we look at the task of directing, because the oldest and most important device for achieving goal coordination is the supervisor.

COACHING SUPERVISORS

The chief duty of a supervisor to his own superior is to see to it that his subordinates are coordinating goals among themselves and with other groups. The supervisor doesn't directly coordinate the work of his subordinates in all instances. The supervisor also must employ directional devices, teach principles of coordination, illustrate their application, and apply tests to determine the quality of synchronized effort. Getting supervisors to accomplish these tasks is the responsibility of the maintenance manager. Figure 23-2 lists five rules of goal coordination that have proven to be effective. They should be applied by all maintenance supervisors, including the manager of the department.

FIVE RULES FOR ACHIEVING GOAL COORDINATION

1. Make sure goals, plans, and instructions are clear and comprehended.
2. Communicate freely and clearly.
3. Make sure subordinates know the relationship between their activities and those of others inside and outside the maintenance department.
4. Stress the need for communications and coordination both inside and outside the maintenance department and the need for keeping superiors informed.
5. Use different techniques for different people and different situations.

FIGURE 23-2

RULES FOR COORDINATION

GETTING OFF TO AN EARLY START

Coordination of goals is most effective if it is invoked in the early stages of planning. It becomes more difficult to properly unify and time departmental plans after they have been implemented. Unity and timing are best achieved when all affected subordinates participate in the planning phase. There is plenty of evidence that a lower quality of effort results from a "tell them nothing" attitude. Management direction should effect this coordination.

ORIENTING

One of the key elements in directing is to provide subordinates with the information necessary for intelligent action. Obviously, the more an employee knows about his work and its environment the more intelligently he or she can work.

ORIENTING NEW EMPLOYEES

Part of the function of directing is to give the new employee information essential to his or her assignment—a detailed explanation of the new employee's job and its relationship to other activities.

The job itself needs to be clearly described with respect to its purpose, scope, and the authority delegated. A written position description helps this process along. The new employee should also be advised as to how his performance is to be evaluated.

Organization and reporting hierarchies should also be explained to the new employee, together with his relationship to other subordinates. An organization chart facilitates this process.

Finally, the new employee should be personally introduced to employees within and outside the maintenance department with whom he will have a lot of contact.

All of the above items take a considerable amount of time to accomplish, but they are essential to proper directing. A personnel department may be able to orient a new employee to the parent organization and cover some of the details of employment such as the fringe benefit program, but it is up to the new employee's immediate superior to handle all of these other items. Not only is it his responsibility to orient the new employees, the immediate superior is also the person best able to do the job. Doing it also gives the superior one additional advantage—it immediately exposes the subordinate to his leadership. Further, doing this part of orientation himself, and seeing to it that his supervisors also do it properly, can head off a lot of grief, both for the new employees and for the manager.

CONTINUING ORIENTATION

Often orientation is limited to strictly a new employee. It should not be. There is a need for a constant briefing of all employees on immediate and continuing departmental activities, orientation to new assignments and changes in the activities of the parent organization. The difficulty in practicing continuous orientation of employees is that the job is never finished. A manager has to be at it constantly, and it is easy to grow indifferent to an activity that must be repeated indefinitely.

There is another important need that is satisfied by continuous orientation. The best coordination of goals occurs when individuals see how their jobs contribute to the major goals of the maintenance department and the parent organization. The individuals need to know and understand these goals. They learn about the goals largely through continuous orientation.

ORIENTING YOUR SUPERIOR

A third facet of orientation is your boss. You have a responsibility to keep him informed, even if it is only to give him information that he needs to keep out of trouble. The part that takes real intestinal fortitude is telling him when your operations have gone awry. But you might as well tell him, because he is going to find out eventually anyway. You are probably doing yourself a favor by letting him hear it from you.

SUPERVISING DIRECTLY

You cannot direct people effectively in a maintenance department without having face-to-face contact. This type of contact is sometimes called direct supervision or personal observation.

Through direct personal contact a manager is better able to communicate, to coach, and to expose his subordinates to his leadership. More importantly, it is through face-to-face contact that he can get a feel of operations, receive suggestions, and receive other information that he could not obtain otherwise. Delegating tasks is one thing, and keeping in touch with performance is another.

Let's look at a classic example of the violation of direct supervision.

A new manager for a railroad maintenance yard was appointed from another district. When he would come to work, he would go directly to his office. He never said hello to anyone and never left his office until quitting time. He even ate his lunch in his office. Many of the workers and front-line foremen did not even know what he looked like, let alone what he wanted done. He purposely tried to limit his personal contacts to his clerk. The three general foremen who reported directly to him seldom saw him. They usually got all their instructions in written memo form. They were supposed to submit any questions in writing, and a reply would be made in writing.

This super-recluse thought that he was doing a good job as yard manager. An internal audit revealed otherwise. Three key indications were that productivity had dropped 35 percent below other maintenance yards, labor grievances were 20 percent higher, and 40 percent of the supervisors had requested transfers while he held the job as manager.

The railroad sent in one of its happiness boys—a consulting industrial psychologist. After a skillful analysis, the consultant came to the conclusion that the yard manager ought to spend more time talking to people. After a year of trying to get him to talk to his men, it was finally decided that he was just too introverted. So the railroad sent him back to his old staff job and put in a new yard manager.

Our super-recluse was well-educated, intelligent, and knowledgeable in the field of railroad maintenance, but he was a dismal failure as a manager. He purposely avoided direct supervision.

DIRECTING WITH UNITY OF COMMAND

The technique of unity of command has long been used in organizing. In essence, this technique consists of subordinates being responsible to only one superior. This style of command precludes division of loyalties and conflicting orders and priorities.

Unity of command is also used in directing. Directing is carried out most efficiently by one person. There should be no outside interference in the supervision of subordinates. The immediate superior should know better than others the nature of the subordinate, his technical proficiency, and to which motivation he best responds. The immediate superior is in the best position to select whatever directing techniques maximize productivity of the group of subordinates, as well as the individual subordinate.

The facilities maintenance department in a firm in Southern California had a real challenge in making this principle stick. The industrial relations department was aggressively undermining maintenance supervision by encouraging individual employees to come to them. As a result, there was a constant stream of maintenance personnel going over to the industrial relations department on all types of things. Supervisors in the department were being completely bypassed. Industrial relations personnel were trying to make all sorts of decisions ranging from vacation scheduling to performance appraisal. The manager of the maintenance department was getting ready to retire. Despite the complaints from his supervisors, he failed to do anything. But it did not take long for his successor to take some leadership action. He went directly to the manager of the industrial relations department and told him that he had no intention of tolerating this interference in his operations. The industrial relations manager agreed with him and the activities ceased. But the situation had been highly disruptive to operations, because unity of command had been violated.

24 Applying Proven Managerial Techniques in Maintenance Work

Like love and war, managing has certain working rules or principles. However, the mere understanding of the rules will not assure good management. The art of management consists of knowing how and when to use the rules. This fact is particularly obvious in management direction, which deals exclusively with the human factor.

This chapter is important because it covers the proven methods of how and when to use the techniques of direction presented in chapter 23.

USING THE STAFF MEETING

The practice of having a weekly staff meeting is common in maintenance departments. However, there is a marked difference in the quality of these meetings. This difference is largely attributable to a failure on the part of some maintenance managers to use this type of meeting effectively. Used properly, the staff meeting can be an effective means of applying the direction techniques of communicating, leading, orienting, and enhancing coordination of goals.

Staff meetings take up the time of all of the subordinates who attend them, as well as your own time. Make sure this time is productive for all participants. There are three major reasons for having a staff meeting. These are:

- To give out information
- To get information
- To identify challenges

GIVING OUT INFORMATION

The staff meeting provides an excellent opportunity for:

- Defining authority relationships
- Interpreting policies and procedures
- Explaining the reasons for courses of action

RECEIVING INFORMATION

Each subordinate attending the staff meeting should pass on some word about his operations. The staff meeting should be used as a means of encouraging subordinates to integrate their operations and achieve a high quality of goal coordination. The

310

subordinate should be taught to report on things he or his people are doing that have an impact on the operations of other subordinates.

The main purpose in using a staff meeting to get information is to encourage goal coordination. The information reported by subordinates should affect the operations of other subordinates. A big time waster in staff meetings is the subordinate who spends his reporting time moaning about petty problems in his own operation that have no effect on the operations for which his peers are responsible. He should be solving these matters himself or work directly with the maintenance manager to solve them without taking up the time of his peers. It's up to the maintenance manager to exercise leadership and coach subordinates to eliminate this practice.

Another reason that some staff meetings are not productive is that a manager uses the meeting to chastise an individual subordinate or the group as a whole. Naturally, people do not look forward to attending a meeting where they are going to be severely criticized for their individual performances or for their performance as part of a group. Let's look at an example.

The manager of a maintenance department for a manufacturing firm consistently used his staff meetings to criticize the performance of his immediate subordinates. As a result, the attendees never reported much about their operations and were prone to send a subordinate to the meetings. Not giving out information hindered goal coordination and defeated one of the purposes of having a staff meeting. Sending subordinates added another complication. The manager was meting out discipline to his immediate subordinates in front of workers who were lower down in the chain of command. This action undermined the leadership of his immediate subordinates.

Fortunately, the maintenance manager's boss attended one of the staff meetings and observed what was going on. In a coaching session with the maintenance manager he pointed out to him what was happening. The staff meetings got back on the right track.

IDENTIFYING CHALLENGES

One of the most fruitful things that can come out of a staff meeting is the identification of challenges. As each subordinate reports on his operations, matters often come to light that need resolution. The key to having a truly productive staff meeting is to identify those items and assign responsibility for working them, *but do not work them out at the meeting.* A big time waster in staff meetings is to have two subordinates discussing a challenge that affects their operations but has no bearing on the operations of other peers. This sort of discussion wastes the time of the men who do not have a share in the challenge and changes the staff meeting into a problem-working session. Working out problems in a staff meeting is usually the primary reason that many of them are of such low quality and maintenance supervisors lack enthusiasm about attending them. It is up to the maintenance manager to prevent this problem from arising by pinpointing what the challenge is, assigning the appropriate subordinates to work on resolving it, and moving the meeting on to other topics. Here is an example.

In the maintenance department of a school district, the staff meetings lasted anywhere from three to four hours because the manager used the staff meeting for problem-working. Problem-working in staff meetings often took up the time of three or four of his immediate subordinates who were not involved with the challenge, and turned the meetings into very long sessions.

Then the manager began to use his staff meetings only to identify challenges rather than to work them out. The result was that the staff meetings began to require only one hour. At the same time, each subordinate spent more time reporting on his operations. Thus, the manager got more information, goal coordination was enhanced, and the meetings were no longer an endurance contest.

TIMING

The timing of the staff meeting is important. Three things must be taken into consideration when picking a time for a meeting.

First, pick out a time when your people can attend without disrupting their operations. For example, avoid trying to hold the staff meeting when it will interfere with a shift change. Any foreman worth his salt wants to be out on the floor directing the work between the shifts. He doesn't want to be tied up in a staff meeting.

Second, select a time that is conducive to making the meeting productive. In one maintenance department, the manager tried to hold his meetings on a Friday afternoon. He could not have picked a worse time. The foremen were tired at the end of the week and more concerned about getting home on time. The atmosphere was not conducive to getting information, receiving information, or identifying challenges. Further, nothing would be done about a challenge until the following Monday unless the manager made them work on a problem on the weekend. The foremen were not too keen about problem solving on their days off, and they tended to hold back anything that the manager might think ought to be worked out immediately. When the manager changed the time to Monday afternoons, the staff meetings became much more productive.

Third, pick a time that coordinates with your superior's meeting. This coordination can be set up for giving information or for receiving it, depending upon which one has the highest priority. Let's look at some examples.

One maintenance manager holds his staff meeting on Monday afternoons and then goes to his superior's staff meeting on Tuesday mornings. This enables him to field any questions on the status of the maintenance department. His main interest is in having the most current information on departmental operations available for his superior's meeting.

Another manager takes a completely different tact. He places a high priority on using his staff meeting to give out information. He schedules his staff meeting to immediately follow his superior's meeting. He has the most recent information to pass on to his subordinates about the parent organization. "I want to make sure that they get the word from me, and not through the grapevine. Sometimes the word gets garbled in the grapevine and some of my guys see shadows that aren't really there."

HAVING AN AGENDA

Staff meetings invariably are more productive if there is an established agenda. An agenda permits each subordinate to think about what he wants to report and to be aware of topics that will be covered by others. An agenda disciplines the flow of the meeting. It also adds a uniqueness and a stated purpose to each meeting, which help keep the staff meeting from becoming a ritual instead of a purposeful gathering.

USING MINUTES

Preparation of minutes to a staff meeting can have several advantages. These are:

1. Minutes provide a written historical record of what went on in the meeting. This history can prove helpful in checking when certain decisions were made or information distributed.
2. The minutes can be an effective device in documenting action items assigned—particularly those concerned with working on identified challenges. These action items can then be reviewed at the next meeting for progress reports.
3. The minutes can be used to enhance communications. Circulation of copies of the minutes provides a means of communicating staff meeting activities to both employees of the department who did not attend and the maintenance manager's immediate superior.

APPLYING THE TWO-TIER STAFF MEETING

Some maintenance managers like to hold two-tier staff meetings. A two-tier staff meeting is one in which the manager has his immediate subordinates and their immediate subordinates attending the meeting.

BENEFITS

There are several advantages to the two-tier meeting. These are:

1. It gives the manager some face-to-face contact with the next tier of supervision and exposes them to his leadership.
2. It can encourage goal coordination by having the second-tier managers get a better understanding of how their group affects other operations and what other operations are doing that affects them.

PITFALLS

There can be drawbacks if the two-tier method is not used correctly. Here are some pitfalls.

1. The maintenance manager must guard against violating unity of command by issuing orders or suggestions directly to a second-tier subordinate. This action would bypass his immediate subordinate. In one department, the immediate subordinates began to absent themselves from the meetings and send their subordinates to the monthly two-tier staff meeting. This situation should not be allowed to occur, because the chain of command will get short-circuited when the immediate subordinates are not present. The maintenance manager is bound to begin to assign action items directly to second-tier subordinates. If the practice goes too far, the authority delegated to the immediate subordinate and his leadership will be undermined. In addition, the maintenance manager's immediate subordinate will not be getting the information he should have to do his job more intelligently.

2. Two-tier meetings tie up a lot of supervisors. Proper timing of the meeting can be even more important than that of a single-tier staff meeting if the technique is to be productive. Normally, maintenance managers who use two-tier meetings do so once a month and have a single-tier meeting during the other weeks in the month. This procedure ties up fewer supervisors at least part of the time.

USING 5WH

Facilities maintenance departments are often faced with major construction of rehabilitation programs or relocations to other facilities, which can have a marked effect on the entire operation of the parent organization. Getting the work done is usually identified as a project, and it takes true direction to accomplish it successfully.

One of the biggest mistakes in managing a project is to say when you are going to do something before you have determined how you are going to do it. Without having first determined the specific tasks required to attain the project objectives, the estimated time required to do the job cannot be established with much validity. Unfortunately, the required completion date for a construction or relocation project is often dictated by management. The maintenance department then has to back into a schedule to support the need date.

One maintenance manager for a multisite organization who has been highly successful in handling many of these types of projects uses a style of direction that he calls the 5WH method. The 5Ws are who, what, when, where, and why. The H stands for how. Essentially, the 5WH method applies the techniques related to a participative style of direction. The manager describes his 5WH method this way.

"What we are going to do, why we are doing it, and where it has to be done are usually dictated by upper management. Then it's up to me to determine who is going to do it. After that, its up to the doers to determine how they are going to do it and when they are going to do those things required to meet the need date. In the initial project briefing, I give my management team the what, where, why, and who. They create the how and when. To complete a project successfully, the support and commitment of every subordinate who is going to have to make a contribution to meeting the deadline are required. This commitment is best obtained by allowing them to participate in the planning. One of my tenets of management is that people tend to carry out their own plans more enthusiastically than those made for them. Besides, we usually have to handle six or seven major construction or rehabilitation projects and a couple of relocations every year. I couldn't plan and handle the day-to-day details of getting them all done even if I wanted to."

To make the 5WH method work properly, project performance standards must be agreed upon ahead of time by the manager as well as the subordinates. The establishment of performance standards permits each subordinate to gauge his performance as he proceeds. Control then rests with each performer. In this manner, the subordinate knows when his performance is not acceptable or when his delegated part of the project has gotten out of control. He can then advise his manager that he needs help to put the project back on the right track.

In essence, this approach is management by exception. The subordinate only reports to the manager when there is a deviation from the plan which he needs help in correcting. Then the manager has time to take action to bring the project under control again.

The 5WH approach makes use of delegation. Subordinates will support what they help to create. Delegation enables subordinates to help create, since they do the planning. It also provides for getting off to an early start in achieving goal coordination.

HANDLING PROBLEM-WORKING SESSIONS

Problem-working meetings should be used as a deliberate effort on the part of the manager to bring the people concerned with the subject at hand into personal contact. The purpose of problem-working sessions is not to pass on information, but to encourage workers to integrate their efforts in finding a solution to the challenge and to achieve a high quality of goal coordination.

Sometimes problem-working sessions ramble, and when they are over nothing appears to have been accomplished. This problem is largely attributable to a failure on the part of the manager to demonstrate leadership. How and when do you execute this leadership role? There are eight steps that have proven to be effective. These are listed in Figure 24-1.

In reference to Step 2 in Figure 24-1, letting each member express what he thinks the problem is, some additional comment should be made. It should be expected that members will sometimes define symptoms rather than the real problem. This mistake is not harmful, because when all of the symptoms are laid out, they may help to bring the real problem into focus in Step 3. For example, a construction project may be behind schedule and the whole job is being paced by one craft. This is a symptom. The foreman of the craft pacing the job may state that his problem is that materials are not available. Again, a symptom. The real problem may then be identified as being a weakness in the material procurement system. The solution is identified as being a matter of revising the ordering procedures.

STEPS IN PROBLEM-WORKING SESSIONS

1. Define the challenge.

2. Have each member give his definition. This procedure provides a means for the correct definition of the challenge and a better understanding of its effect on the various operations for which participants are responsible.

3. Reduce the problem to writing. If it can't be reduced to writing, it isn't correctly defined yet. Try again. Don't define symptoms, define the real problem causing the symptoms.

4. Get each member to state what he thinks should be done about it: what action he should take, and what actions are required by other group members.

5. Examine the pros and cons of each course of action.

6. Select the solution that appears most viable and valid.

7. Assign specific responsibilities to participants who are to execute the solution.

8. Reduce the planned solution to writing. If you can't write down the solution and who is to do what, you haven't solved the challenge.

FIGURE 24-1

TENETS OF PROBLEM-WORKING

When using the eight steps listed in Figure 24-1, there is another important item to remember. You must stay with the challenge at hand. If others come up, write them down. But sift them out for separate action, otherwise you end up with a laundry list of everyone's gripes and no solutions to any of them.

Problem-working sessions can be highly counterproductive if they are permitted to degenerate into character assassinations. The manager should execute his leadership to prevent this problem from occurring. The members should be there to seek a positive and lasting solution—not to participate in a rock-throwing contest.

TOURING

The use of on-site tours is an effective means of direct supervision and face-to-face contact, of communication, and of exposing the subordinate to your leadership style. It is particularly useful when the maintenance department has personnel at off-site locations.

Subordinates should welcome tours—not fear them. A positive attitude on their part is achieved by making them feel that they get something out of showing you around. Further, if they have a positive attitude about your being there, they are more apt to open up and communicate their concerns.

There are certain hows and whens to follow to conduct tours successfully. First, make sure you have allowed sufficient time to make the tour and to spend some time with the subordinate. If you rush through the tour, you are lessening your chances of learning something because the subordinate does not have the opportunity to tell you everything that he may want to talk about. This lack of communication defeats one of the major reasons for taking the tour.

Second, avoid making any criticisms during the tour. If there are things that need to be corrected, save your comments until the tour is over. Bring them up after you have reviewed the things that appear to be running right. If you criticize during the tour, two things are apt to occur. First, the subordinate is quite apt to become defensive and stop being open with you, which would nullify some of the reasons for taking the tour. Second, some of the subordinate's crew may overhear the criticism, which could result in an undermining of his leadership.

Lastly, when the tour is over, always give the subordinate some feedback. If you're inscrutable, he won't know where he is, how he is doing, what's being done right, and what can be done better. He has a right to know, or you should not have given him the job in the first place. Further, you will not demonstrate any leadership if you haven't anything to say.

COACHING

The purpose of a coaching session is to provide time devoted to working specifically with the individual subordinate on a face-to-face basis. The intention should be to provide a means of implementing the direction requirements of motivating, communicating, leading, teaching delegation, orienting, and supervising directly.

Coaching or review sessions should be geared towards reviewing the status of operations or programs assigned to the immediate subordinate—how things are proceeding, what has been done, and what is to be accomplished in the future. Coaching sessions are an excellent method of keeping in touch with performance.

WORKING THE WHEN

There are three key elements in facilitating the when aspect of coaching sessions. These are:

- Schedule the sessions.
- Allocate enough time.
- Shield the sessions.

Scheduling Sessions

Some managers fail to use the coaching or review session technique effectively because they do not make a concerted effort to formally schedule them. They rely upon the subordinate to take the action of requesting a meeting or let the sessions occur on a spur-of-the-moment basis. Putting the responsibility on the subordinate to initiate coaching sessions is a failure to exert leadership. Relying on spur-of-the-moment sessions weakens the use of the technique because there is no purposeful planning before the session occurs.

Coaching or review sessions that are scheduled for a specific time each week or month provide for some advance planning, both on your part and on the part of the subordinate. Keeping in touch casually may be advocated by some managers and theorists, but if you allow your coaching sessions to occur by chance some subordinates are apt to be missed, particularly where the maintenance manager has a multisite operation and some subordinates are geographically isolated.

Allocating Time

Sessions that have to be broken off because of other meetings or commitments are seldom as productive as those that are permitted to run their full course. How much time should be allocated for a given session is largely governed by the subordinate. Veteran subordinates who have been trained in the techniques of completed staff work and exception usually require substantially less time than the subordinate in a new position. You have to get a feel for how much time each subordinate will normally require. One maintenance manager found that his coaching sessions ranged from twenty minutes to an hour and a half, depending upon the individual involved. He allocated his time accordingly.

Shielding the Session

To make a coaching session productive, the meeting should be shielded from interruptions. Do not allow phone calls or people to break into the session and destroy its continuity. Many managers prefer to conduct the sessions in their own office where they can assure that there will be no interruptions. Others prefer to use the subordinate's office, where they feel the subordinate is apt to feel more comfortable and data that may be required will be more accessible. If you use the subordinate's office, make sure he knows to shield the meeting.

WORKING THE HOW

Although each manager develops his own style, there are four tenets to be observed when conducting coaching sessions. These are:

- Gauge the session to the individual.
- Communicate.
- Let the subordinate take the lead.
- Take action.

Although some of the above items may appear to be obvious, it is amazing how often they are not practiced.

Gauging to the Individual

Two things should be taken into consideration when gauging a session. First, the need for and extent of coaching will vary markedly between the seasoned veteran and a person who is new in a position. Hopefully, the veteran is familiar with policies, procedures, and organizational alignments. There should be less need to use a coaching session for this purpose in dealing with him. The session, however, still provides an opportunity for continuous orientation of the veteran.

Second, individual differences in personalities should be recognized. The coaching session is used most successfully for motivating by those managers who understand their subordinates as individuals and follow their native instinct in dealing with them. If you tend to look at your subordinates as a group rather than as individuals, they may very well respond uniformly. Subordinates are a potential source of ideas, inspiration, and creative effort. The value of a subordinate lies in what he can contribute to his own areas of assigned responsibility, and also in the creativity he can bring to bear on other areas. You will not get this value from each subordinate in an atmosphere of uniformity.

Communicating

By definition, a communication is an exchange of information. A conversation requires a dialogue, not a monologue. There is little value to a coaching session that becomes your monologue, because you are not getting anything out of the subordinate. By the same token, if it becomes the subordinate's monologue, he is not getting anything from you. While each participant has to learn to listen, he must also learn to contribute if there is to be any true communication.

Letting the Subordinate Lead

Letting the subordinate take the lead is the third tenet of successful coaching. It is his session, so let him tell you what he thinks is important about his operations or program and where he needs your help. When he is finished you can cover the items you planned to discuss. Better still, work your items into the conversation when they are related to something he has brought up.

Letting the subordinate lead facilitates getting a feel for operations, receiving suggestions, and getting other information that you might not otherwise obtain. It also permits you to get some insight into the subordinate's frame of reference. You will discover what is frustrating him and what judgments he is making about his section and about the work. You will also find clues as you talk with him as to how in touch he is with his subordinates and how he views his subordinates and their work. This knowledge can lead directly to some determinations on what he needs to do better and where he might need some orientation or training in delegation.

Taking Action

To demonstrate leadership, you have to take action on those matters where the subordinate indicates that he needs your help. Tell him what you intend to do and *make sure you do it*. Nothing destroys motivation in subordinates faster than a superior who refuses to exercise leadership or does not live up to his promises. Future coaching sessions are apt to be of little value if the subordinate gets the feeling that you are not really concerned about him or the operations and programs for which he is responsible. There is no substitute for genuine interest. The authenticity of your interest is demonstrated by your taking action.

APPLYING THE TECHNIQUE OF EXCEPTION

As stated in the previous chapter, the technique of exception requires delegation of authority. The subordinate only refers those matters that extend past the scope of his authority to the superior. He only goes to his superior when there are challenges with which he needs help.

Proper execution of this technique requires the avoidance of certain pitfalls. Its application must be tempered by the need to keep in touch with performance, which is why touring and coaching sessions are so important. If you are not aware of the warning signals that come before the exceptions occur, it may be too late to help the subordinate before operations become substandard.

There is another snare. Under the guise of the technique of exception, a maintenance manager can spend most of his time and energy with the underachiever because he needs the most help. This situation is unfair to your other subordinates, who are being neglected when you don't give them an equal amount of your time. Further, it can deflate morale. Subordinates who are pulling their share of the load are quite apt to wonder why they should continue to do so when you seem to be ignoring their efforts and spending all your time with the subordinates who are failing to properly use the authority that you have delegated to them.

The main purpose of the technique of exception is to delegate authority and avoid the details of a subordinate's work. But the manager is not absolved of his responsibility for concentrating on morale. If you never go to a subordinate or hear from him unless challenges arise, you are not truly directing.

APPLYING THE TECHNIQUE OF COMPLETED STAFF WORK

At best, maintenance operations are usually hectic. There is little opportunity for a maintenance manager to play the role of a corporate executive and initiate extensive, time-consuming studies before making all decisions. Administering day-to-day operations requires rapid decision making, because time is of the essence. However, the maintenance manager still has some opportunities to apply the technique of completed staff work. It is applied most successfully in planning and on special projects where time is available to develop a solution.

Planning can be directed towards devising better methods of organizing, staffing or controlling. Special operational projects may involve such items as a major relocation, a major construction or rehabilitation program, setting up an off-site maintenance operation, establishing a scheduled maintenance program or any other new operational project within a given maintenance department.

As stated in the previous chapter, the key to successful application of the technique of completed staff work is to make sure the subordinate does his own thinking. But in having the subordinate do his own thinking there is frequently a fine line between giving him guidance and giving him the answers.

When the technique of completed staff work is applied, it supports several other tenets of direction. First, the individual essentially has to develop a plan or a solution. This participation generally leads to commitment, because people tend to support what they have participated in creating. Second, it can supply motivation by giving the individual the opportunity to apply special talents and exercise his capacities, which constitutes job enhancement or fulfillment.

HAVING A DEPARTMENTAL MANUAL

One of the major reasons for having a manual is to support principles of management identified with managerial directing. As indicated in Figure 24-2, the specific tenets supported are the enhancement of goal coordination, the availability of policies and procedures, the orientation of new employees, and the transmittal of information.

Some large maintenance departments, particularly those involved with multisite operations, have found that a departmental manual is beneficial. However, the quality of these manuals varies. This difference is attributable to the fact that one or more requirements for having a useful manual have not been observed in preparing and maintaining the manual.

A manual is a tool to help people remember. It is a place to preserve systems documents so that employees can get the information when they need it. A useful manual must have meaning and usefulness for the people who use it. To achieve usefulness, it must be written from the viewpoint of the user. A truly useful manual is a reader's manual. It is not written for the benefit of the writer or to satisfy the whims of some executive.

HOW A MANUAL SUPPORTS THE
MANAGEMENT FUNCTION OF DIRECTION

1. TO FACILITATE COORDINATION

The manual provides a place to record a decision on how to do work. Putting the decision into writing helps everyone in the department to know how to work together and with other departments.

Recording the decision in writing and putting it in a manual where it is available to concerned personnel should preclude the necessity for having to make the same decision over and over again.

The manual should achieve better coordination by eliminating inconsistencies. By writing down exactly how each section in the maintenance department takes part in getting the work performed, you will provide better consistency. Inconsistencies generally cause grief with department employees, other employees of the parent organization, and vendors. If you do not reduce your policies and procedures to writing you will be prone to inconsistency.

2. TO MAKE POLICIES AND PROCEDURES AVAILABLE

If they are going to use it, maintenance department personnel should have policy and procedural information readily available. Also, the fact that you have reduced them to writing will cause you to look at your policies and procedures critically and make sure that they make sense.

3. TO ORIENT NEW EMPLOYEES

The manual can be a valuable tool for orienting and training new employees. By reading the manual the new employee learns how to do his job faster and how the various sections of the department interact.

4. TO ENHANCE THE TRANSMITTAL OF INFORMATION

When you write down your instructions on how to do work, you eliminate many incorrect interpretations that you would probably get if you relied on word-of-mouth transmittal.

FIGURE 24-2

REASONS FOR HAVING A MANUAL

PLANNING A MANUAL

If it is to support managerial direction, a useful maintenance department manual should be structured like a systems manual. Figure 24-3 lists the five elements that such a manual should contain.

CONTENTS OF A MAINTENANCE DEPARTMENT MANUAL

1. Statements of policy to guide future actions.

2. Procedures that affect the various sections of the maintenance department.

3. Organization charts, including a top chart of the major section of the maintenance department, and one showing how the department fits into the parent organization.

4. Descriptions of the functions or responsibilities of the various sections within the maintenance department.

5. An index that truly facilitates finding information in the manual.

FIGURE 24-3

CONTENTS OF A MANUAL

The first step to take in planning a useful manual is to determine the information needs of your department. What is to be put into the manual depends upon who is going to get it. The information needed is directly related to the distribution plan. Some other key considerations in planning the manual are:

1. Prepare the documents in a uniform style in order to make it easier for the reader to understand the information.
2. Keep headings for the sections of the manual simple and direct so that they clearly define the documents. Call the various documents what they really are—e.g., policy, procedure, organization chart, and section responsibilities.
3. Have each document not only contain a descriptive heading, but also have a place for a date, a number and a space to indicate what the document cancels when it is sent out as a revision.
4. Do not plan to prepare infinitely detailed procedures if the recipients of the manual do not need them.

PLANNING MANUAL COSTS

To create and maintain a manual costs money—more money than most managers think. As a rule, you can figure that, on the average, a seasoned procedures analyst is going to take about forty hours to develop a single procedure. Thus, if you figure that your planned manual will probably contain about fifty procedures, a minimum of 2000 hours will be needed, which is one man-year of procedure analyst time just for the procedures section of the manual. In addition, there will be the costs for binders and reproduction costs for the required number of copies of each document in the manual. There should be enough binders and copies of documents to take care of the initial

```
┌─────────────────────────────────────────────────────────────────┐
│                    MANUAL DEVELOPMENT COSTS                       │
│                                                                   │
│  Labor                                                            │
│                                                                   │
│        Procedure Analysts                  $12,000                │
│           (12 man-months @ $1000/month)                           │
│                                                                   │
│        Typing and Proofing                   3,600      $15,600   │
│           (6 man-months @ $600/month)                             │
│                                                                   │
│  Materials                                                        │
│                                                                   │
│        Manual Binders                      $   100                │
│           (50 @ $2 each)                                          │
│                                                                   │
│        Preprinted Formats and Dividers         150          250   │
│                                                                   │
│  Reproduction Costs                                               │
│                                                                   │
│        Organization Charts                 $    20                │
│           (10 @ $2 each)                                          │
│                                                                   │
│        Procedures                            1,250        1,270   │
│           (50 manuals w/50 procedures                             │
│            @ an average of 5 pages each                           │
│            w/cost per page of 10¢)                                │
│                                                                   │
│  Total Development Cost                                  $17,120  │
│                                                                   │
│                         FIGURE 24-4                               │
└─────────────────────────────────────────────────────────────────┘
```

FIGURE 24-4

SAMPLE MANUAL COST FORMAT

distribution list, plus some back-up binders and copies in case the distribution list is expanded. Figure 24-4 is a format that can be followed in estimating the costs of creating a manual.

As another rule, you can figure that a minimum of 10 percent of the manual will require revision annually. If this maintenance is not performed, the data in the manual will become obsolete rapidly. This obsolescence will render the manual useless and negate the very purpose for having one in the first place. Therefore, if you are going to have a manual, be prepared to incur the annual costs necessary to maintain it properly. Let's look at an example of what happens if you do not maintain a manual.

To facilitate direction, a large maintenance department responsible for multisite research and manufacturing facilities decided to have a manual. The manual cost $40,000 to develop over a period of nine months. At first it did prove valuable, particularly in coordinating off-site maintenance section activities and orienting new personnel while the department was experiencing a growth of about 20 percent per year. However, when the firm ceased to grow and some cutbacks occurred, the department ceased to maintain the manual. Within three years, the manual was no longer being used since much of the data had been rendered obsolete by organization changes. The $40,000 investment had gone down the drain because the department had failed to spend $500 a year on manual maintenance over the three-year period. When company growth began again, the manual had to be virtually rewritten in its entirety to make it useful. This project cost another $30,000.

This example indicates two rules that should be closely followed if your maintenance department is going to have a manual. First, be prepared to incur the maintenance costs on the manual in order to keep it useful. Second, do not try to defer this maintenance, because the delay will cost you more money in the long run.

EVALUATING AN EXISTING MANUAL

If you already have a manual and want to see if it is truly useful, find out by a field survey. Take a sampling in the various sections of the maintenance department and ask the following key questions.

1. Who is using the manual?
2. How often is it used?
3. Can users find the information when they want it?
4. Is the information written clearly and concisely?
5. How can the manual be improved?

After the survey you can take action to improve the usefulness of the manual.

GETTING THE WORD TO ALL EMPLOYEES

The dissemination of information to employees at the worker level is a part of managerial direction. For example, a bulletin from the parent organization announcing what will be the paid holidays in the coming year or a departmental bulletin on the policy of borrowing tool crib items for personal use must be transmitted to the workers. Selection of the method to use in transmitting the information is dependent upon how the maintenance department is organized and individual preference.

USING THE SUPERVISOR

To support the chain of command and unity of command, some maintenance departments prefer that all information in either verbal or written form be transmitted through the lines of supervision.

The front-line foreman, in turn, has several methods that he can use to pass on the information. One of the most widely used methods is the use of the foreman's shift change briefings to transmit information. At these briefings, copies of organization or department bulletins are distributed, or other information is given verbally. This method has two advantages. First, it provides an opportunity for any feedback or questions that might arise and permits the foreman to provide any required clarification.

Second, the shift change briefing is an established time for all hands reporting to each foreman to be in attendance. The word can be passed without disrupting operations or incurring lost production time by calling a special meeting.

USING OTHER METHODS

Some maintenance managers do not feel that it is necessary to use the chain of command in all instances. They feel that general information to be disseminated to all employees in the form of a department or organization bulletin can be distributed by other means without unduly hampering unity of command. Many of them use bulletin boards. The bulletin board may not always be totally effective as a means of transmitting information. If men report directly to a job site, they may never get to where the bulletin board is located. If they do, they are probably there on other business and will not take the time to scan the board.

Distributing a copy of the memorandum to each concerned employee with his paycheck or with his time card is an alternate method. Some maintenance departments put correspondence in the time card slot where the employee keeps his time card. This way they don't have to wait until the next time the paychecks or time cards are distributed to pass out the information.

Some maintenance departments still use the all-hands meeting to disseminate information. Usually these meetings are held at shift change time so as not to disrupt operations too much. In large maintenance departments or those organized on a zone basis, the all-hands meeting is not always effective for two reasons. First, you have to gather all of the various crews together from their normal reporting places. This procedure results in lost time. Second, there is usually that certain percentage who don't get the word and fail to show up. These people have to be briefed separately.

Managers of smaller maintenance departments where all men report in at a single place can use the all-hands meeting effectively. Some managers prefer to use the all-hands briefing. They feel that giving the briefing themselves provides an opportunity for all departmental employees to be exposed to their leadership.

INDEX